Training
with Video

Steve R. Cartwright

Knowledge Industry Publications, Inc.

White Plains, NY and London

Video Bookshelf

Training with Video

Library of Congress Cataloging-in-Publication Data

Cartwright, Steve R.
  Training with video.
  Bibliography: p.
  Includes index.
  1. Television in technical education. I. Title.
T65.5.T4C37   1986   371.3′358   86-2902
ISBN 0-86729-132-X

Printed in the United States of America

10   9   8   7   6   5   4

Contents

List of Figures

iv

Preface

This book is a natural extension of the seminar "Designing and Producing Video Training Programs" that I have been presenting over the last three years. The questions and responses that I have received from participants have encouraged me to expand the information presented in the session into a book about the design and production process in greater detail than I can cover in the seminar.

My experience as a trainer and video producer has formed the basis of this book. Extensive interviews with numerous trainers and video managers have expanded the production processes and provided further examples. And often I cite numbers that represent industry standards.

My goal in writing this book was to create a reference source that trainers and video producers can employ to justify the use of video in training and to provide a guide for the design and production of video training programs. The book provides a practical approach to video training program development. It is not meant to be a study of adult learning theory but to prepare the reader to start working on effective training programs.

The ten chapters in this book offer approaches to video training design and production with emphasis on the program *design* in Chapter 4 and video production *planning* in Chapters 5, 6, 7 and 8. Chapters 1 and 2 offer a source for the use and justification for video in training, and Chapter 3 offers budget considerations. Chapter 9 details what resources and equipment are needed to create video training programs. Chapter 10 offers actual case studies of successful video facilities that produce video training programs. This chapter provides a valuable reference for production budgets, facility and staff size and special uses of video for training.

TRAINING WITH VIDEO

I am grateful to the following reviewers for their many valuable comments and suggestions: Bill Hight, District Manager, Video Development, AT&T Resource Management Corp; Bill Van Nostran, Van Nostran and Associates; Lon McQuillin, president, McQ Productions; and Jerry Freund, Director of Education, Tucson Medical Center. I would also like to express my appreciation to Jim O'Hara of the Tucson Police Department, Jim Barefoot of Belk Stores, Alex Tice of Union Pacific Railroad, Steve Kultala of Adolph Coors and Jerry Freund for providing the case studies. I would also like to thank Bill Marriott, Dick Handshaw and my brother, Phil Cartwright, for numerous conversations that clarified my ideas for and approaches to the shaping of this book, and Ellen Lazer, my editor at Knowledge Industry Publications, which provides the forum for the seminars that started this book.

Finally, I would like to thank my wife Karen for her patience and support in typing and reviewing the many drafts that appeared on the computer screen.

November 1985
Tucson, Arizona

For Christopher and Kevin

1

Why Video?

Television's effectiveness as a tool for training has been demonstrated at every level of education, from preschool through adult education.[1] Its use for training in industry, government, health care and education has increased dramatically in the last decade. It shifts time, space and content into viewer-adjusted, viewer-accepted segments, making information available whenever, wherever and in whatever amounts needed. It uses color, motion and sound to convey technical skills, concepts and attitude equally well. Because it is a popular medium that is proving itself as an effective training method, television is influencing the way we train significantly.

This book discusses how television has become such a successful training tool. It examines why, how and when you can use television in training. It describes a successful approach to creating effective video training programs. It details the design, scripting, production and editing processes. This is a "how-to" book, researched and written for one purpose: to help you get results through video training.

THE ADVANTAGES OF USING VIDEO

But first a look at why video is being used as a training tool, why corporate video departments produce more television programs each year than all three major networks combined, and why corporations, hospitals and government agencies spend millions of dollars a year to purchase equipment and video.[2]

1

Video is an unusually successful training tool. It saves time, money and resources. It adapts to a wide variety of situations and requirements, and at the same time, it guarantees consistency in repeated viewing unmatched by the live performance. It makes use of the inherently interesting features of color, motion and sound in one universally acceptable medium.

How Video Saves Time

Even though it usually takes about ten times as long to create a video training program as it does to prepare and present traditional classroom training, video ultimately saves far more time at the point of delivery and when the material must be repeated. With a well-designed video program, you save up to 75% of the time it would take to deliver the same message in traditional lecture style. If training time can be reduced by as much as 75%, money and other resources can certainly be saved as well.[3,4]

At Sperry Corporation's Semiconductor Operations, for example, the training manager is Neil Decker. He asserts that training time was reduced by 50% as soon as he started to use training modules that incorporated video. This meant that the operators of technical equipment who participated in the training became productive in half the time. He expects this training cycle time to be reduced even further as his trainers become more efficient with the modules.

In designing traditional media materials like slides and transparencies, you have the advantage of compression. You can cut out unnecessary information and deliver only as much information as is directly related to the objective. This also holds true in video program design. When you compress information, the student can learn in one-third to one-fourth the time it would usually take in a classroom setting, so the objective of the training situation can be obtained in much less teaching and learning time.[5]

At the Pima County Sheriff's Department in Tucson, AZ, where video is used as a training delivery method, what once took an hour of instruction is now delivered in a 20-minute tape. The Department redesigns hour-long training sessions into video programs and delivers them on videocassette to the deputies in the field. This takes the deputies off the patrol beat for much less time.

Not only does the use of video in training save student time, it also saves instructor time. No longer tied up in the classroom, instructors can devote more time to program development. This translates into more information being delivered to the student.

How Video Saves Money

By reducing the time spent by staff, management and instructors, video can reduce the cost of training and communications. If it takes less time to communicate corporate messages through the use of video, staff and management time is saved; if it takes less time to train new workers, they become productive that much sooner; if the training process itself takes less time, instructional staffs may be reduced, or their time may be employed more economically. Thus, video effectively creates cost savings throughout the corporation.[6,7]

Alexander Tice, Media Production Supervisor of the Union Pacific Railroad, finds that basic job skills can be taught much more easily and cost-effectively through video: "We have new employees watch a videotape and they soon have a pretty good understanding of what is required of them to complete the job. This does save us a lot of time."

C. J. Silas, Chairman and Chief Executive Officer of Phillips Petroleum Company, believes that video saves the corporation money: "I'm impressed by the way that the wise use of video can save money by eliminating some expenses, travel costs and employee time."

Silas also comments on the effectiveness of video as a communications tool:

Video is an important facet of an effective communications program for us. I like the sense of immediacy, emotion and action that our video crew captures in its tapes. They've shown us that video, whether it's used for training, corporate news or management communications, can go beyond teaching facts and procedures. It can help motivate people and show them that they are part of Phillips's success, and I believe that makes a difference in a company's performance.

How Video Increases Efficiency

Seldom is the viewer ready to receive information at the same time the trainer is ready to deliver it. Video enables you to bring these two times together and allows the learner to adjust information reception to his or her own schedule. No longer are you locked into prescheduled information delivery. The learner now controls the schedule of viewing.

This is particularly useful for organizations like hospitals, law enforcement agencies, transportation and high technology manufacturing that require workers to be scheduled round the clock. At the Tucson Police Department, Officer Jim O'Hara explains:

> I put out a videotape. . .and in three days the entire department
> has seen it. Anyone who was on sick leave watches the video
> when he or she gets back. This takes the pressure off instructors
> to deliver training round the clock trying to meet schedules.

With well-designed video training, instructors do not have to be present to deliver information. Thus, video can assume the responsibility of training 24 hours a day. A typical example of using video as a training tool on a 24-hour work basis can be observed at the Tucson Medical Center. This hospital has set up an effective distribution system for its training tapes through the use of video carts, master antennae, a closed circuit system and satellite reception. Programming is available around the clock to patients and staff. Hospital clinical skills, equipment operation, patient information and surgical techniques are made available through the various distribution systems.

How Video Reduces Travel Time and Costs

Time and money are saved when video replaces the traveling instructor. Using a video network (video players placed in field offices) learners in regional or international offices can benefit from on-site training without the instructors having to travel to each location.

This is demonstrated dramatically at Bank of America in San Francisco where, according to Robert Ripley, Assistant Vice-President and Executive Producer of Media Services:

> We've got 2000 offices spread out across the country. If people
> need to be updated or trained on a particular subject, we
> use video training tapes. We can reach up to 80,000 employees

worldwide with one tape, saving travel costs to receive that message.

Video extends the viewing audience and opens up communications among local, regional, national and international offices with the use of video players and monitors.

At Burr-Brown Corporation, a leading manufacturer of microelectronic components in Tucson, AZ, training is transmitted to sales offices around the world by means of videotape. Thousands of travel dollars have been saved by using video as a delivery method. Sales staff do not have to travel to Tucson to receive new product information. Product information and demonstrations are sent to sales offices worldwide via video.

This is also true for the Adolph Coors Company in Golden, CO. A network of 600 VHS players serves its distributors all across the country. Programs are sent out for sales training, product information and corporate communications. Its success is measured in terms of the sales results created by the tapes and the money saved from travel.

Video Training Is Consistent

Another significant advantage of using video in training is its ability to deliver information consistently.[8,9] All viewers receive the same information in the same style of delivery. The videotape records the lecture or program and preserves its style and meaning exactly as the presenter originally intended them. Individual differences of a number of instructors in the interpretation, style or philosophy is never a problem. All viewers receive the same message every time they view the tape.

This is extremely important when it comes to teaching attitudes and philosophy, as in supervisory or management training. The presentation of the organization's policy and philosophy is not left to individual interpretation. Instructors need not feel the pressure of presenting the material over and over again in a consistent, organized manner. All viewers, whether they are impressionable new employees receiving their first exposure to the company or seasoned sales personnel in offices around the country, receive consistent information—information that has been designed, written and controlled by the organization. Video creates a reliable, constant delivery system, readily available, delivering the same corporate message each time.

An example of this is cited by Scott Carlberg, Supervisor of Video Communications at Phillips Petroleum:

Our chairman, Pete Syless, wanted to talk to everyone in the supervisory classes about his philosophy of performance. He obviously couldn't go to all the classes, so we recorded him. It works well because the participants are hearing corporate philosophy from the top man.

Video Is Inherently Interesting

Video is a visual medium, and it has the ability to hold interest through color, motion and sound. If used properly, video can create interest through visual stimuli.[10,11,12] According to one expert, visual images hold interest, and add to retention and recall of information.[13] Visuals affect accuracy and standardization, as does color.[14] Color helps illustrate difficult concepts and clarify information. Various studies[15] have proved that color and visuals can accelerate learning by 78%, and improve and increase comprehension by 70%. They can increase recognition and perception, and reduce errors.[16] Color and visuals affect motivation and participation. Video can turn the viewing situation into a positive, motivating experience.

Video, like film, has another advantage that contributes to training, and this is motion. Through motion you can provide performance feedback, role modeling, product and safety demonstrations, process examples and motion skill development.[17,18] Motion helps to deliver the message. If used effectively, it increases the training reception. In some situations, such as skill development, motion is essential for effective learning. Officer Jim O'Hara reports:

We are using the effects of motion with video more and more for our training. In firearms training, we are showing the proper way to approach the target but we are also showing details of trigger motion and control: how to squeeze the trigger smoothly, even down to the proper way to draw the weapon. We have to have the ability of motion to demonstrate these techniques.

Audio is an element of video that helps in many training situations as well. You can use audio to reproduce various voice and sound conditions. You can also use it for recognition and discrimination. Sound effects, music and voice add interest, realism and excitement. Audio in video productions is frequently overlooked, but if incorporated into the script and into the objectives, audio can add to any training experience.

DISADVANTAGES AND LIMITATIONS OF VIDEO

Although the advantages of using video as a training tool far outweigh its disadvantages, you need to consider certain drawbacks before you purchase the equipment or start the video production process.

Cost

The cost of producing video programs is high in comparison to that of traditional classroom approaches to training. Slides, audiotapes and transparencies are generally cheaper to produce, and require less time for preparation. The overall production of videotapes is time-consuming. It generally takes ten times as many hours to produce a good video training program as it does to produce traditional audiovisual aids for classroom use.

Other Limitations

- Type size and amount of type needed to deliver the message on screen. Because of video's rather low resolution characteristics, the size and amount of type can become a problem on the small TV screen.
- Availability of credible on-screen talent. Video programs require good performers who have credibility with the audience.
- Fast turnaround time requirements may not allow for lengthy video production preparation.
- The skills required to write a script, direct the talent, and light and edit a program may not be readily available.
- Delivery of the program may be difficult. If a playback network is not in place corporation-wide, viewing the tapes may become a problem.

The playback of video programs to audiences sometimes becomes complicated if the organization has distant facilities or offices and does not have control over the actual playback of tapes. Delivery is often overlooked as you rush into the excitement of getting into video. Organizations need to define delivery means, responsibility and maintenance clearly. The best produced tape with a strong message is useless if no one has the means to see it.

SPECIAL FEATURES OF VIDEO

Standard Format

One of the outstanding advantages of video is the ease with which it can deliver information because of the standardization of video formats. In the early days of video, several different formats were not compatible with each other. A tape created on one system, or standard, could not be played on another, different system. With the introduction of the 3/4-inch U-matic videocassette by Sony Corporation in the early 1970s, however, the compatibility problem has been greatly diminished. This format quickly became the standard for production and playback in the industrial videotape industry. The ease of using a videocassette, rather than reel-to-reel machines, and the standardization of the format help make the delivery of video an easy operation. In later chapters we will cover in detail the production and distribution equipment needed in training situations.

The most popular production format currently used for recording industrial training programs is the 3/4-inch U-matic videocassette, and the leading format for distribution in the United States is the 1/2-inch VHS consumer format.[19,20]

Portability of Equipment

Most of the money spent on video production equipment is now going for single-camera systems that can produce videotapes in the field. Portable video production equipment now becoming available at reasonable prices offers producers the realism that field shooting provides. It helps video managers get away from the high cost of maintaining studios and allows them the freedom of shooting more productions on location.

Field production creates realism for the training situation. It allows you to visit the work site and present the working situation that the trainee will experience eventually. It gets away from the stale, standard talking head of studio production and lets the viewer travel visually to the sites and to the experiences that are needed to deliver accurate training information. Field productions bring in new scenes and experiences that could not be created on videotape before.

New Emphasis on Post-Production

Since producers are shooting more footage in the field in a single-camera mode, they are spending more time in the post-production process. They are shooting on 3/4-inch U-matic videocassettes and coming back to the facility and editing the videotapes. Video producers are spending more money on editing equipment that can handle the needs of field production.

Trends in Distribution

Distribution is taking a slightly different direction. Training and video departments are turning more frequently to a smaller format for delivery—the 1/2-inch videotape. Marketing and sales surveys indicate that training distribution networks, like the consumer market, are utilizing 1/2-inch VHS more frequently than any other distribution system.[21,22] This smaller tape format saves tape costs and playback machine costs. Communications and training tapes are making their way into the home for viewing at night and on weekends; VHS, the more popular consumer format, allows for a broadening of the audience for this purpose. With VHS distribution we are expanding our training capabilities beyond the classroom. Videotape offers ease of transferring from one format to another. Tapes produced on 3/4-inch U-matic can easily be transferred to 1/2-inch VHS.

Videocassettes and videodiscs linked by way of computer interfaces are becoming more popular with trainers today. Videodiscs offer fast access time to information and work well in interactive systems. The interface allows for interactive training to take place. Coupled with the microcomputer, discs allow speed, accuracy and low-cost distribution for high-volume video training. The personal computer interfaced with videocassette or disc will become a popular, efficient training tool.

VHS, videodisc, computer-enhanced video and videoconferencing are distribution systems that we will look at in more detail in Chapter 9.

NOTES

1. T. R. Ide, *Media and Symbols: The Forms of Expression, Communication, and Education* (Chicago: The National Society for the Study of Education, 1974), p. 335.

2. John H. Barwick & Dr. Stewart Kranz, *Why Video?* (New York: Sony Corporation of America, 1975), p. 8.

3. Barwick and Kranz, *Why Video?*, p. 8.

4. Eugene Marlow, *Managing the Corporate Media Center* (White Plains, NY: Knowledge Industry Publications, Inc., 1981), p. 73.

5. Walt Robson, *Video: A Corporate Picture* (New York: Sony Corporation of America, 1978), p. 8.

6. Marlow, *Managing the Corporate Media Center*, p. 73.

7. Paula Dranov, Louise Moore and Adrienne Hickey, *Video in the 80s* (White Plains, NY: Knowledge Industry Publications, Inc., 1980), p. 35.

8. Dranov, Moore and Hickey, *Video in the 80s*, p. 35.

9. Ronald H. Anderson, *Selecting and Developing Media for Instruction* (New York: Van Nostrand Reinhold Co., 1976), p. 73.

10. Greg Kearsley, *Training and Technology: A Handbook for HRD Professionals*, (Reading, MA: Addison-Wesley Publishing Co., 1984), p. 29.

11. Jerrold E. Kemp, *Planning and Producing Audiovisual Materials*, (New York: Harper & Row, 1980), p. 21.

12. Kemp, *Planning and Producing Audiovisual Materials*, p. 22.

13. Francis M. Dwyer, *Strategies for Improving Visual Learning* (State College, PA: Learning Services, 1978), p. 12.

14. Dwyer, *Strategies for Improving Visual Learning*, p. 12.

15. Ronald E. Green, "Communicating with Color," *Audio Visual Communications*, November 1978, pp. 14-47.

16. Dwyer, *Strategies for Improving Visual Learning*, p. 150.

17. Anderson, *Selecting and Developing Media for Instruction*, p. 73.

18. Robert Heinich, Michael Molenda and James D. Russell, *Instructional Media and the New Technologies of Instruction* (New York: John Wiley & Sons, 1985), p. 206.

19. Judith M. Brush and Douglas P. Brush, *Private Television Communications: Into the Eighties* (Berkeley Heights, NJ: International Television Association, 1981), p. 89.

20. Dranov, Moore and Hickey, *Video in the 80s*, p. 4.

21. Ibid.

22. "Max Troubles for Betamax," *Time*, January 16, 1984, p. 60.

2

Applications of Video

An estimated 10,000 companies in the United States use video for a variety of communications and training applications.[1] Among the most popular uses of video in industry, according to a number of surveys, are those of basic employee skill training, advanced employee skill training, and management and supervisory training.[2,3,4] In addition, the use of video in sales and customer training also deserves mention in this chapter.

BASIC SKILL TRAINING

In basic employee skill training, for example, clerical and operator-level training, video plays an important role. Video's capabilities are well-suited for teaching clerical skills, telephone selling techniques, work flow processes and innovations.

Video works well in this kind of training because it can show a physical process in step-by-step fashion and with visual and verbal detail. In manufacturing, basic skills such as machine operation and maintenance, test procedures, product handling, quality control and product testing are all taught effectively with video.

Basic manual skill training often takes the form of an overview of the skill being taught followed by a series of specific learning points. Often an employee manual or workbook accompanies the videotape. By viewing the videotape before working through the manual, employees can be speedily prepared to perform that particular skill process. Problems in a manufacturing process can be identified and shown to people for

13

training purposes by attaching video cameras to microscopes and testing equipment.

In the microelectronics industry video training programs are used to increase what is called yield. In manufacturing, output = % of input. That means that the finished products equal a certain percentage of all of the materials, labor and waste that went into the manufacture of those products. This percentage is the yield. For example, when input is 100% and you come out with 95% product, your yield is 95%. Obviously, the manufacturer wants to create the highest yield possible. The yield is affected by such factors as scrappage, breakage, mishandling, reworking and the building process. (Fig. 2.1)

My experience producing video training programs for Burr-Brown Corporation, IBM, Sperry Corporation, Northern Telecom and General Instruments has shown me that video training programs can affect yield positively. The training programs Burr-Brown produces are referred to as modules and consist of videotapes, instructor's guides, student manuals, slides and training audits. The modules in effect train operators on the correct use of specific manufacturing equipment. They emphasize safety precautions and operations theory. The modules cover the entire semiconductor manufacturing process—from wafer fabrication and assembly to packaging.

Our data show that video training has increased yield by as much as 20%. Depending on the size of the manufacturing plant, this percentage can represent millions of dollars per month. Video's consistency of training (see Chapter 1) guarantees and controls the manufacturing process, which in turn controls yield. The goal in manufacturing is to gain control so that everyone involved in the manufacturing process will follow the same step the same way each time. These training programs also affect cycle time—the amount of time it takes a product to go through the manufacturing process from start to finish.

Videotape can also be used to evaluate a learner's new skill as it is being learned and after the learning has taken place. For example, after going through the learning experience, the operator may be required to perform the new skill in front of a camera for recording on videotape. The videotape performance can then be viewed and evaluated by the trainer. This evaluation process is being used in law-enforcement training with excellent results.

At the Tucson Police Department, Officer Jim O'Hara uses videotape to evaluate the approach police officers make on routine traffic stops:

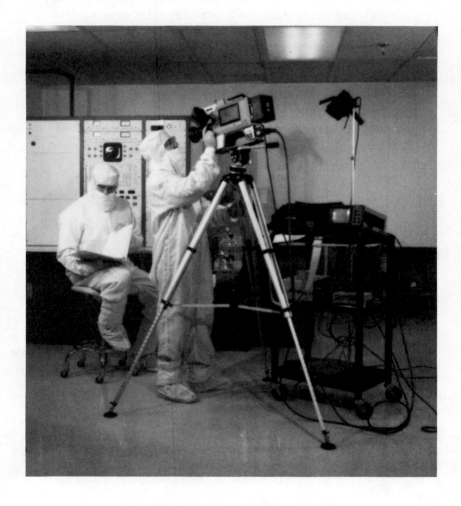

Fig. 2.1. Author on location in clean room environment, shooting video training program for microelectronics industry.

People have adverse reactions when being approached by police officers with their hands on their guns. We have seen officers making routine stops with their hands nonchalantly resting on their guns. This looks sloppy and becomes intimidating to the public. We videotape these same officers making a stop during training exercises and we tell the officers that they did not make a good approach and that they had their hands on their guns. The officers usually will not admit to doing so and quite often the moves are actually made unconsciously. We play back the tape and the officers see their mistakes and it turns into a real learning experience.

Another way we use video is for driver training. In vehicle dynamics, it's extremely useful. You can show the students how the car reacts in curves and how the tires react on both dry and wet surfaces. One of the areas we had trouble with was trying to teach people how to go through a slalom course, and how to move the steering wheel smoothly and not jerk it. They didn't understand, even after watching the instructor drive through it. So we put a camera behind the instructor and focused on different actions. We combined this action with the reaction of the vehicle and the tires and put the whole thing together to show the cause and effect.

Modeling

Modeling a process to acquire a new skill is educationally sound according to many research findings. Video is like film in that with its attributes of repeatability and motion, it can teach skills through modeling the process.[5,6] This technique proves effective in teaching procedures like chemical handling, safety instructions around machinery and simple computer operation. With the process and skill modeled on videotape, the operator can see the correct way to do the job and be able to duplicate the performance soon after viewing the tape. (Fig. 2.2)

ADVANCED SKILL TRAINING

Advanced employee "people" skills, such as interview techniques, public speaking, listening, customer relations and supervisory skills, and high-level operational skills, such as computer operation, computer software

Fig. 2.2. A Phillips Petroleum Co. video crew shoots a scene from a training program on removing asbestos safely.

usage, and high technology equipment maintenance and repair, can all be taught with videotape. Interviewing and counseling techniques can be modeled on videotape for the employee to watch and repeat. In public speaking, listening and customer relations, training tapes provide not only modeling but also feedback by recording and playing back the way the employee responds to a particular problem or how an employee handles a particular customer.

Wilson Learning Corporation, a successful Minneapolis-based producer of training materials, incorporates video into its *Customer Contact Curriculum* by providing sensitive customer relations problems on videotape. Linda Antone, Vice-President of Video Production, says that Wilson Learning is devoted to video because of its effectiveness in illustrating and demonstrating concepts. She states that their training design always includes a visual component. "Our videotapes carry actual teaching messages supported by print. . . . We feel video is an effective asset to our programs because of its powerful impact."

Wilson Learning Corporation provides numerous customer relations, management and supervisory programs, all using video as an integral part of their training packages.

At Tucson Medical Center, Jerry Freund uses video for performance feedback and role modeling. In a program on performance evaluation he videotapes the participants conducting performance evaluations and then plays back the tapes for critiquing. "The tape is extremely effective in allowing the participants to see their performances. Hearing a critique from an instructor is one thing but actually seeing your own performance is quite another." (Fig. 2.3)

SUPERVISORY AND MANAGEMENT TRAINING

Positive role modeling is valuable in the education process. It increases enthusiasm and interest in the learning situation, and increases retention of the material.[7,8] At the supervisory and management training level, videotape provides a positive role modeling situation. Trainees can watch the proper techniques for handling grievance procedures, dealing with a problem employee or establishing a reward mechanism. Then they can role-play a vignette or a scene, videotape it and learn by watching their own mistakes.

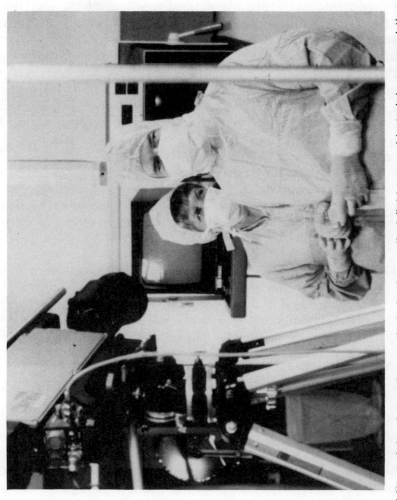

Fig. 2.3. Video training is used extensively in the medical field to provide training around the clock. The portability of video equipment allows the producer to record "on-scene" in the operating room.

Management training often uses videotape for speech training and the development of interpersonal skills. Management's role lends itself largely to dealing with people skills. With role-playing and problem-solving scenarios, video can contribute to management training.

A vignette, such as that of a manager dealing with a problem supervisor, can be played out on videotape. This way a consistent role-play is presented each time the class takes place. The use of professional actors or talent who are not identifiable on videotape can save the embarrassment of peer review in classroom role-playing situations. A scene is played before a manager, and the manager is asked to react to it. The proper way of handling the scenario is played back on videotape so a manager can see a problem, work with that problem his or her own way, and then see the best way to play that scene if it were to come up in activities during the day.

At CRM/McGraw-Hill Training Systems the successful *Supervision Series* uses behavior modeling through videotape to teach management and communication skills. Video can show effective supervisors in action. Trainees observe and analyze the key skills being portrayed on video. CRM/McGraw-Hill literature on the series explains:

> The video models are carefully structured with clearly defined skill points, making the model easy to follow and practice. Because the models are credible and brief, participants will concentrate on technique, without unnecessary distractions.
>
> The behavior modeling technique is based on progressive learning. Research shows that it is easier and more effective for trainees to learn skills on a relatively simple level first, practice them, and then apply them to more complex situations (p. 2).

EMPLOYEE ORIENTATION

Another major use of video is new employee orientation. This presents corporate structure, employee benefits, services and perhaps a welcome by the president or executive officer to a new employee. Videotape creates a consistent message, ready to be delivered when and where the orientation takes place. A well-produced program creates a positive image of the company for new employees. This can be a lasting impression reinforced with good training at the beginning of careers. Scott Carlberg at Phillips Petroleum explains their use of video for employee orientation:

> Remember when everyone was just getting into video and they did their employee benefit tapes and their audiences slept through them? Our approach is a lot different now. We really break things down into small pieces of information and tell employees the good news about the plans as opposed to fill out form A and form B. The employees appreciate the information and it becomes more useful for them. (Fig. 2.4)

SALES TRAINING

Companies throughout the United States are finding that technical sales training tapes can increase sales. Videotape is often used in sales training to improve sales technique and presentation style. Videotape can also be used to send information about a product out to the sales force. The better informed the sales staff can be about a product, the better they will be able to sell it.

Burr-Brown Corporation produces sales training programs to provide product knowledge, applications, marketing strategy and manufacturing background to a worldwide sales staff. The marketing department conservatively estimates that these sales tapes have increased total sales of a particular product by 8% to 10%. This results in a gain of approximately $100,000 per year, per product.

At T.G. & Y. stores, a major discount retailer headquartered in Oklahoma City, video training brings positive sales results. According to Marcy Treadway, Supervisor of Employee Development, their customer relations sales training programs create "remarkable results." At T.G. & Y., customer relations are stressed. The programs the company produces teach basic job-related skills, but they also emphasize customer relations. With 755 stores serving 27 states and employing 27,000 people, the family discount store has to provide sales training that creates results economically. Video training has helped them meet the challenge. According to Treadway, "Our video programs work well because they do not leave room for interpretation; and in customer relations, sending out a consistent message that follows the philosophy of the company is important."

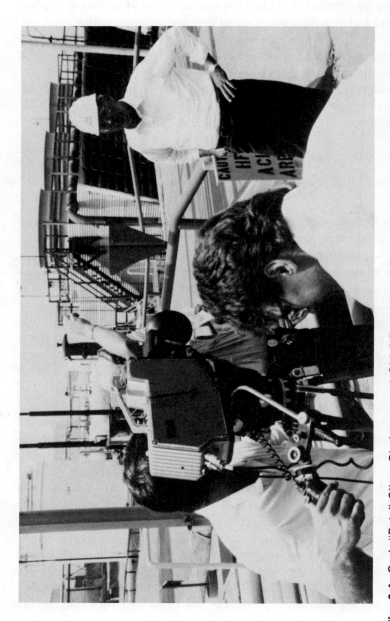

Fig. 2.4. C. J. "Pete" Silas, Chairman and Chief Executive Officer of Philips Petroleum Co., tapes his on-camera portion of a training program about corporate safety on location at a Phillips refinery in northern Texas.

They have concluded that it is cheaper to send out their sales training tapes than it is to bring store managers in for training. They have traced direct sales increases to the use of the videotapes. So at this large discount store video is not only saving money, but it is making money for the corporation.

CUSTOMER TRAINING

According to *Training Magazine's Industry Report, 1984*, a yearly survey of training trends in the industry, customer education has been the greatest growth area over the last three years, and video plays an important role in reaching the customer. Customer education is "a highly cost-effective method of promoting product awareness, increasing customer satisfaction and lowering technical service cost."[9]

I foresee videotape playing an ever-increasing role in the delivery of customer education, especially in the introduction of newly purchased products. For example, in high-tech industry, where a machine may cost a million dollars, the customer expects the company to provide training and follow-up information on new applications and maintenance procedures. Video permits the customer to train all new personnel, on any schedule, and to continue to provide training long after the sales representative has gone.

According to Eugene Marlow, President of Media Enterprises, Inc., a New York-based company specializing in product information and marketing videotapes, the single most influential factor for the growth of customer education is the new portable playback equipment. With VHS and monitor packed in a briefcase, the sales rep can walk into an office and present a graphic demonstration of the product to a potential customer. Marlow predicts that with increasing sales of home video players (over 27 million players in the home by the end of 1985, according to Consumer Electronics Association) a whole new market for customer education will be created. Merchandisers can give product demonstration tapes to potential customers and ask them to watch the tapes at home at their convenience.

Belk Stores, a Southern retail chain, calls customer education videotapes a good investment. Jim Barefoot, their video manager, says that through "customer-education videotapes that are made available in our stores, we sell more products and create excitement about the products that the tapes are selling."

Barefoot goes on to say that "customer education is a tremendous growth area that is paying off for us. We are creating 'why-buy tapes,' two-minute educational tapes on a particular product that explain to the customer the value of the product, why it is a good product to buy from the customer's point of view. These tapes are being accepted very favorably in each of our stores that have video playback. They are working well."

SAFETY TRAINING

Safety is a priority subject in health care, law enforcement and industry. For example, videotapes are being produced to show proper chemical handling and the effects that the chemical has on the product as well as the user. Videotape dramatically models safety situations and drives important safety points home. M.E.E.S., Inc., of Tucson, AZ, distributes a videotape series called *Safe Handling of Non-Flammable Ethylene Oxide*, designed for the health care and industrial market. The program describes the importance of safety in handling ethylene oxide and portrays safety procedures that should be followed in the event of a major leak or spill.

In law enforcement, videotape is being used to model proper procedures and safety techniques for firearms training. At the Tucson Police Department, the video unit produces safety training tapes for police officers in the field. Recently, when the department implemented a new firearms procedure that required officers to fire, holster and move from position to position with loaded weapons, the new firing sequence introduced serious safety considerations. A videotape that would be shown to all officers through their training video network outlined the new safety rules and introduced the officers to the requirements of the course. The program succeeded in alerting the officers to the new safety measures and alleviating any apprehension they might have about the new course. The adoption of the new sequence of firing went smoothly with no accidents. (Fig. 2.5)

Alexander Tice of Union Pacific Railroad speaks highly of the use of video for safety training: "We use video as part of an overall approach to getting the message across. We use video to analyze problems in locations where people really can't work. We set up the camera to record in dangerous settings to see what is going on so it can be analyzed safely, later."

Fig. 2.5. Officer Jim O'Hara of the Tucson Police Dept. edits a video training program on firearms safety.

If the training calls for a change in skills, behavior or attitude; if it is safety, management or technical; or if the audience is staff, management or customers, video training creates results. Video is a flexible, efficient medium capable of handling a myriad of training situations.

NOTES

1. Paula Dranov, Louise Moore and Adrienne Hickey, *Video In The 80s* (White Plains, NY: Knowledge Industry Publications, Inc., 1980), p. 13.

2. Judith M. Brush and Douglas P. Brush, *Private Television Communications: Into the Eighties* (Berkeley Heights, NJ: International Television Association, 1981), p. 55.

3. Dranov, Moore and Hickey, *Video In The 80s*, p. 23.

4. "Training Use Stable," *Video Manager*, September 1985, p. 1.

5. Robert Heinich, Michael Molenda and James D. Russell, *Instructional Media and the New Technologies of Instruction* (New York: John Wiley & Sons, 1985), p. 207.

6. W. R. Daniels, "How To Make and Evaluate Video Models," *Training and Development Journal*, December 1981, pp. 31-33.

7. Robinson P. Rigg, *Audiovisual Aids and Techniques* (London: Hamish Hamilton, 1969), pp. 15, 16.

8. Rigg, *Audiovisual Aids and Techniques,* pp. 11, 12.

9. Ron Zeiwce, "Customer Education: The Silent Revolution," *Training*, January 1985, p. 26.

3

The Cost of Producing Video Training Programs

1986 dollars low for Detroit area because of auto industry

Television costs a great deal of money to produce, but the cost of producing video training programs must be measured against end results. The average in-house produced video training program costs an organization $6000, an accepted industry standard. That may seem high, but if all costs are considered, the figure of $6000 is believable. I will break down this cost figure later in the chapter.

There are as many ways to determine the costs of a program as there are organizations using video, so costs usually vary slightly from one company to the next. But several accepted formulas can be used as guidelines for this study: cost per program, cost per viewing minute and cost per employee.

COST PER PROGRAM

Determining the cost per program is one of the easiest ways of getting a fairly accurate picture. An organization's total video operations cost is divided by the number of programs produced in one year.

A major part of total program cost goes for staff, which includes annual salaries, benefits and staff expenses. Doubling actual staff salaries provides a good estimate of the total personnel costs to the organization (including all benefits and support services, such as accounting, personnel, secretarial time, etc.).

27

A typical operating budget for our study would be $80,000. This direct expense budget would include salaries for two people (including benefits), travel, office supplies, equipment repair, educational seminars, and telephone and equipment amortization. It would not include indirect expenses for building rent or utilities because overhead varies from organization to organization depending on the size of the facility. But many video facilities break down their overhead into what it costs per square foot in utilities to keep the space in operation. This overhead cost would then have to be figured into the operating budget.

Other costs that may be included in the operating budget are additional staff time devoted to the project, graphics, photography and outside services. All of these costs can be lumped into a yearly operating budget.

The cost per program can be an attractively low figure if the organization is producing many programs. But if it costs the organization $80,000 a year to maintain its video production facility, and you produce only two programs a year, the cost per program is $40,000. If the organization produces from 12 to 15 programs per year, the per-program cost will be considerably less.

This formula represents an average program cost because all programs do not cost the same amount. A simple demonstration tape or someone speaking into the camera delivering information costs much less to produce than a program that is fully scripted and shot on location. As a result, the average program cost can be a little misleading, but it is a useful guideline.

COST PER VIEWING MINUTE

The cost per viewing minute, sometimes referred to as cost per finished minute, can be determined by taking the final program running time and dividing that into the program costs. For in-house videotape production a reasonable average cost per minute to produce a fairly sophisticated, scripted program is approximately $500 to $1000 per finished minute. This kind of program, a "Level Two" program, is described in Chapter 3. The costs are generally determined by how much staff time is used to produce the program plus operating expenses. To determine the dollar value of staff time you divide the number of available working hours into the total staff budget or determine an average hourly figure for services. An accepted average for the amount of working hours available in a year per person is 1680. That is derived by multiplying

40 hours x 52 (2080) and subtracting 400 hours for vacations, holidays and breaks. That leaves 1680 chargeable hours per employee.

In a moment we will look at the average time it takes to produce a training program. These times can then be broken down into hourly costs. Direct and indirect expenses also have to be figured into the hourly rate or the finished minute formula.

COST PER EMPLOYEE OR VIEWER

In determining the cost per employee or viewer, we simply divide the total program cost by how many viewers will actually see the program. If our program costs $6000 to produce and 500 people will view it, the cost to the organization is $12 per viewer. This can be a very attractive formula to use in justifying video if a large audience will be seeing a program. If a program will cost from $7 to $12 per employee in a fairly large organization, that cost is often easy to justify.

You can take this one step further and calculate what effect that $12 will have in value to the organization. Will that better-informed employee produce more after viewing the program and can a dollar figure to be placed on that production?

A helpful budgeting resource is *Practical AV/Video Budgeting*, by Richard E. Van Deusen, published by Knowledge Industry Publications, Inc.

LEVELS OF PRODUCTION

Since there are so many different types of video training programs and each has different levels of involvement, it helps to create categories of program sophistication. These categories will also help in budgeting. One program may take more design time; one program may need more graphics; one program may use professional talent, etc. These and other considerations have to be figured into the program costs. Generally, programs may be broken down into the following four categories:

Level One

Level One is usually a program that does not require a script, such as a simple demonstration or talking head; a program that does not require any type of preparation, graphics, editing or outside field production. An example of this type of program would be a content expert delivering a presentation, which would be recorded for others to view if they could not attend the live lecture. Basically the camera is turned on and the lecture is recorded live.

Level Two

Level Two is the level most frequently employed in organizations beginning in video. It uses an outline script or partial script and a content expert to deliver the information. The program may be a demonstration with some slides or graphics. It would require a minimum of editing and preparation. Generally, information tapes, equipment demonstrations or training procedures are found at this level. An example would be a new product introduction. The marketing director would stand in front of the camera, demonstrate the new product and refer to some slides or graphs. This level would not require much preparation and would usually be shot in the studio.

Level Three

Level Three is a fully scripted program. It is usually well designed, well written, fully visualized and storyboarded. A Level Three program generally includes studio and field shots. Usually it is narrated with no on-camera talent. Pictures are edited to the audio narration. A professional talent or someone on staff might read the script.

This type of program requires design and preparation time, logistics, creative directing, shooting, lighting and editing. Most of the programs done in the corporate video setting are usually Level Three programs. They usually require graphics, music mixing and careful editing. Many of these programs have support material (workbooks) to accompany them.

An example of a Level Three program would be a technical training tape in which a trainee would learn to operate a piece of equipment by viewing the correct procedure. Theory of operation would also be discussed, and a student manual might be prepared. Another example would be a supervisory training program in which examples of appropriate and inappropriate behavior are shown through well-designed role modeling.

Level Four

The most sophisticated level, this type of program can be found in the commercial market. It is well-designed, fully scripted and usually shot on location or in a studio with complex sets. It usually incorporates a professional on-camera actor. Special effects such as dissolves, wipes and computer graphics are often employed with extensive post-production requirements. A complex sound track is recorded in a studio and often original music is scored. A typical Level Four program might be an important product introduction utilizing a well-known talent or a complex marketing tape involving locations around the country with multiple talent, sophisticated graphics and original music.

REQUIRED PRODUCTION TIMES

The cost of production is closely related to the hours spent in total program development. Often an organization will calculate only the hours spent in production, and that will create an inadequate program cost. Average production times I have listed below are based on industry standards. These average times should provide a guide for the program planning process. The times noted are based on in-house production with a staff producer/writer assigned to the program. In reality most of us wear several hats and are working on several projects at one time. The time periods given below are based on having one person assigned strictly to one program.

The times listed cover the complete development of a hypothetical training program involving research, scriptwriting, location shooting and editing. The program is a Level Two or Three production. A Level One program would require only two or three days to produce.

Research

Researching a program usually takes an average of one to two days. This depends on many factors including your knowledge of the subject and the type of program you are producing.

Research may encompass everything from determining the actual words that will be spoken in the program to uncovering the past history of the subject, current and historical sales figures, future market trends, etc.

Research is also required to determine which people need to be involved in the project, and whether any media have already been produced on the subject.

Design

Design usually takes about three days. The design of the program, as explained in Chapter 4, requires the identification of the problem and audience, what learning strategy will be used, and how the program will be sequenced and paced. Determining the design goals is a sophisticated area of program production and often requires the services of an instructional designer.

Writing

The average time required for writing a program is three to six days. Like the other creative processes, writing may encompass a wide variety of skills and activities. These might include research, design, visualization, storyboarding and interviewing. The end result, of course, should be a complete shooting script.

Producing

This part of program development includes working out the logistics of equipment and staff arrangements, supervising rehearsals, recording voice narration and shooting the actual tape. The average production time is two to three days.

Editing

The average time for editing, two to three days, will vary, of course, depending on the complexity of the program, how many graphic inserts there are, and how many scenes, music mixes and voice tracks there are to edit. Plan on a day's worth of editing for every day of shooting.

Distribution

Duplicating and distributing the program to the viewing audience generally takes about one day. Videotape cannot be high-speed duplicated like audiotape, so a 20-minute program will take at least 20 minutes to duplicate.

Adding up our average times gives us a two-to-three-week period to produce a Level Three in-house video training program. In actuality it would probably take three to four months to product a well-designed, well-written video training program because most of us work on more than one program at a time.

Again, the average times represent only a guideline for in-house video program development. Each program is different, requiring varying amounts of time for research, writing, shooting and editing, but it is useful to understand the average times involved in each category to allow you to gauge development time for your project.

CALCULATING PROGRAM COSTS

By taking the cost formulas previously discussed, you can calculate program costs. You need to place a dollar value on a day's worth of researching, writing, designing, producing, shooting, editing and duplicating. This day rate will vary depending on the skill required. Accepted industry standards for in-house costs are as follows:

activity		hourly rate*
research	=	$15
design	=	$20
scripting	=	$25
producing	=	$100 (two people plus equipment)
editing	=	$60
dubbing	=	$10

*one person unless otherwise noted

Additional Elements That Affect Cost

Other production elements that affect the cost of producing video training programs are as follows:

Talent

Will in-house, nonprofessional talent or professional talent be used? What about special makeup and wardrobe considerations? The average day rate for professional on-camera talent (not a well-known name) is $500.

Director

Will an outside professional director be hired to direct crew, camera, talent and music editing? This average day rate is $600.

Crew

Will a professional crew be hired to handle the technical responsibilities, lighting, camera and equipment? Day rate averages for crew include the camera, recorder and basic light kit at $1500 (3/4 U-Matic recording format).

Tape

Tape costs for shooting, editing and distribution must be calculated. An average figure for 3/4-inch U-Matic videocassettes is $30 an hour.

Shoot

What special equipment (crane, dolly, sets, props, lenses, specialized lighting and sound equipment) will have to be obtained for shooting the tape? These factors may add as much as $1000 a day to your shoot.

Location

The cost of shooting on location differs greatly from that of shooting in a studio. Usually it costs more to shoot on location because of the special equipment required and the time it takes to travel and set up lighting and sound equipment.

Site Survey

For location shooting a site survey is a must. A day rate for a site survey is usually half the director's daily rate.

Music

Will music libraries be used or special music be composed for the project? Music libraries that give you the rights to use the music from their albums vary in price from $1000 a year with unlimited use to $35 each time the music piece is used in the program.

Graphics

Will studio cards, graphics or slides be used? If these graphics have to be designed and produced outside of the organization, they can cost as much as $100 a piece, depending upon complexity.

Special Effects

Special effects are widely available, and the costs vary from simple dissolves and fades to sophisticated computer graphics and animation. Special effects can and often do add thousands of dollars to the cost of the program.

Distribution

Items to be considered as distribution costs are labels, packaging, shipping, booking, inspection, postage and repair charges. Usually a figure of $10 will be added to the individual tape cost per program for this service.

To illustrate, I have costed out two training programs, one a Level Two production and one a Level Four.

Example of a Level Two Program

This program was designed for the sales staff by the marketing department. Its purpose was to inform the sales staff about a new product, give application examples, show a little of the manufacturing process and indicate potential market impact. It was designed as new product information, to reach 50 regional sales offices around the country. The marketing director was the talent. It was shot mostly in the studio, with some location shots to show product applications and the manufacturing process. The program length was 20 minutes. The marketing director showed slides that gave market information, and he demonstrated the product on camera. There was no script because the marketing director was very familiar with the product information. It took two days to shoot and two days to edit, preceded by a day for logistic arrangements. Twenty slides were used in the program, with each slide costing about $25. There was no hourly rate for using the marketing director (it was he who wanted the program after all) and charges were based on the averages outlined above.

Slides	Twenty slides @ $25 a slide	$500.00
Logistics	One $10-an-hour production assistant. If the department producer/manager had worked on logistics, it would have cost $25 an hour.	80.00
Production	A two-day shoot using two people for crew, shooting on 3/4-inch U-Matic with basic lighting and audio equipment. Studio rate and location rate would be same.	1,600.00
Editing	Two days at $60/hr.	960.00
Duplication	Three days	240.00
Tape	50 copies @ $20 a tape.	1,000.00
	TOTAL	$4,380.00

Example of a Level Four Program

This program was designed to acquaint the viewer with the various steps in the manufacture of a semiconductor. The program showed manufacturing processes and applications of semiconductors as they relate to everyday life. The program used an on-camera professional actor as the host and narrator of portions of the program. The shooting was done at several manufacturing sites, but all were located in one city. There were a few studio shots, but most of the program was shot on location. The program was fully scripted and shot on 3/4 U-Matic broadcast quality with a broadcast camera. The entire program was shot single-camera style. The in-house video department produced the program with rentals of outside scripting assistance, equipment, lighting and post-production facilities. The script called for some animated graphic effects, dissolves, wipes, keys and the use of DVO (Digital Video Effects). The program was edited on one-inch video and the running length was 16 minutes.

Research	Two weeks	$1,200.00
Design	Four days	640.00
Scripting	Five days	1,000.00
Logistics	Production assistant for two weeks	800.00
Production #1	Six days rental of recorder and camera with a camera operator	8,000.00
Production #2	Six days in-house production work by a director and assistant	4,800.00
Grip equipment	Rental of all lighting equipment, shot gun microphone, hand boom, dolly, carts and special filters	1,500.00
Audio engineer	Six days @ $150 a day	900.00
Talent	Five days	2,500.00
Audio	Includes recording services, rights, music mix, narration booth and tape	600.00
Post production	Includes ADO effects, one-inch bump-ups and one-inch master tape	10,000.00
Tape	Includes all 3/4-inch U-Matic production tape and protection copies of final program	400.00
Miscellaneous	Includes catering, paint, props, hardware and lumber	300.00
	TOTAL	$32,640.00

CONCLUSION

Television costs a great deal to produce and each program presents its own fine balance between the desired result, your resources and the cost. These three items influence each other and must be weighed against each other. We have to estimate as accurately as possible the real return on investment. There is usually a high level of expectation for the result and limited resources or budget to obtain it. That is where experience, creativity and careful planning come into play.

4

Video Training Program Design— Phase 1

Producing a video training program takes place in a series of three major phases: the design, production and post-production phases, each of which consists of numerous steps. I have listed 42 of these steps on the chart in the Appendix of this book. By far the most important phase in developing a video training program is Phase I—in which the overall plan for delivering the video training experience is formulated.

Training programs are expected to create changes in behavior, attitude or skill. In order to assess your success in meeting the objectives of the program, you must be able to measure the changes that result from the training experience that you create. The design of the program is the plan that you develop to ensure that you meet your goals; the design becomes the road map that you will follow through the development phase. The design should take into consideration an overall training experience that is developed to solve a particular problem. Videotape may be just one solution to that problem, or it may not be a solution at all. But before you begin to plan the solution, you must first accurately identify the problem.

Five major tasks exist within the design phase. They are: problem identification, audience identification, objective setting, learning strategy and media selection. I consider evaluation to be an ongoing process that takes place during each major phase and discuss evaluation during each of these phases.

DEFINING THE PROBLEM

In most companies the decision to produce a videotape is made before the problem is clearly identified. People tend to jump to solutions before defining the problem. But the program producer should go through a series of questions to ensure that the problem can truly be solved by the use of video. Only after in-depth questioning can you make intelligent decisions about whether to create a training tape or not.

Videotape has the distinct advantages listed in Chapter 1. It is not, however, a panacea for all training ills. Unfortunately, because video production capabilities exist within an organization, too often video is selected to solve problems that could be solved less expensively by the use of other training resources. If the program producer is involved early, during the problem identification step, training delivery decisions can begin, and the decision to use video can enter the formula at a more appropriate time.

Problem identification, or needs assessment, should identify a deficiency in performance and indicate a performance standard against which the deficiency may be measured. You must analyze the deficiency in behavior, attitude or skill level, and design the program to correct it and bring it up to the accepted standard. The responsibility for this analysis usually is assumed by the training director, and the design of the program is often developed by an instructional designer. Nevertheless, program producers must also be acquainted with the process being corrected in order to appreciate the training need fully and offer appropriate media solutions.

Before a project begins, ask these four questions to identify the problem and put it in perspective:

1. *Can training correct the problem?* Training may not be the solution to all company problems. You must analyze carefully to see if training is indeed the correct action to take. Better communication between line and supervision through weekly meetings might be a more effective, long-lasting solution, or perhaps a set of new, simplified forms may achieve the desired result. In other words, what appears to be a training requirement may not be. There may be a simpler solution.

2. *Can the audience being trained solve the problem?* Do they have the needed resources? It may be beyond their means to solve the problem, and no amount of training will prepare them or create the resources needed to solve the actual problem.

3. *How are you going to evaluate the results?* A standard should exist, or one should be made against which to measure the results. Since the standard must be understood by the target audience, it should perhaps become a part of the training program as the accepted way of carrying out the new procedure. The audience has to recognize when there is a problem and should be able to relate to it. They need to understand whether the standard has changed, or if they are not measuring up to the existing standard or whether all the previous ways of doing things are changed. Establishing a standard will also help you create the evaluation tool to measure the program's results against the objectives.

4. *How important is the problem to the overall goal of the company?*[1] The development of a training program, whether it be a classroom lecture, workbook, slide program or videotape, is expensive. You should try, therefore, to calculate what the problem is costing the organization. You must decide how big the problem really is and determine just how high a priority it should have. Will the solution or training affect a large number or just a few? How many people, products or clients are going to be affected by this training? What is the extent of the personnel, money and other resources that should be applied to solving this problem? This question must be worked through honestly before the resources are committed. Quite often the size of the problem and the extent of its influence are subjective. What one person may perceive as a big problem may in actuality be a very small problem. In most cases it is worthwhile to get several people, even the target audience, involved in problem identification.

It is important to understand that at this point in the project you are not talking about videotape. You are primarily focusing on the problem itself, how it relates to company goals, how many people it will affect and how you will know if you solve it. Our energy should be devoted to identifying the problem clearly. You have to look at it objectively from all sides so that you can analyze the situation fairly and offer a just solution, whether it be through videotape, slides, a stand-up lecture or a drink with the boss.

At this point, a solution is selected. If the problem or performance deficiency can be solved through training, your next step is a thorough analysis of the audience.

DEFINING THE AUDIENCE

If I had but one question to ask during the design phase, it would be: for whom am I producing this program? The more accurately you define the audience, the more chance you have to create a success. If the training experience is designed to create a change, who are you asking to change? To whom will the training effort, budget and resources be devoted?

The instructional designer should provide a clear audience profile. This profile will dictate the objectives to be written, and it will provide the program producer with valuable information for script development and media selection.

Audience Profile

The profile describes who the audience actually is. It will influence the design and eventually the script of the program. If training is to create a change, then the people it is to change must be identified clearly.

When you form the audience profile some key questions you must consider are:

- How much does the audience already know about the material being presented? Content knowledge will determine if terms have to be defined, preliminary reading is to be assigned, and to what extent assumptions can be made about content.
- Is the audience motivated to watch the program? Motivation may or may not have to be written in. The question, "What's in it for them?" should be asked continually. Does the audience know why they are watching the program and what effect it will have on their jobs? If the program is asking them to change their performance, do they know what the performance standard is?
- What are the age group and education level of the audience? There should be a difference in the way you present the material to an older, "experienced" group and to a younger, newer group.

This may also influence your choice of music and talent, as well as the way you will illustrate your presentation.

- How large is the audience? Our emphasis so far has been on video programs, but video may not be the best media choice if the training will be presented to large audiences in a large room. We know that television is a personal, one-on-one medium that works well when the viewer is relatively close to the monitor, alone or in a small group, and exercising a certain amount of control over the viewing experience. If the program is to be presented to large audiences, perhaps slides or film, with their superior image quality on large screen projection, would be a better media choice.
- How will the audience see the program? This question is often overlooked. The equipment logistics must be worked out in advance to ensure ease of training delivery. Will the audience be expected to roll the cart out of the closet and turn on the video player, or will this be taken care of? There is no guarantee that the best produced program will be seen if viewing logistics are not worked out in advance. This applies especially to branch offices located around the country. Who will be responsible for showing the program to the intended audience? Distribution must be set up in advance.

Disciplines merge at this stage in the development of training programs. The instructional designer is analyzing the audience so that an educationally sound program can be created to bring about the desired change. At the same time, the program producer is also becoming heavily involved in audience research, so that a meaningful script can be written. In smaller organizations, these two jobs are often held by the same person. Whatever the case, it is important that the designer and the program producer get involved with both problem identification and audience analysis at a very early stage in the development of the program. Many programs that fail to reach their objectives have not merged these two disciplines early enough. Too often, the video producer is handed a script to produce that was written by an instructional designer who has sound educational principles but lacks good visual production principles. Each discipline contributes to the goal. Educational values and production values must be incorporated into one experience.

The importance of audience identification was brought home to me very impressively at a recent scripting session. My colleagues and I were designing a program that was to give the audience an overview of the hybrid technology manufacturing process, that is, the manufacturing of multiple chip devices. It became extremely important that we agree on the intended audience. Was the program going to be shown to engineers who basically knew the design and manufacturing steps involved? If it were, then we could start at a deeper level and go right into the sophistication of the circuits or the process of making the circuits. If, on the contrary, it were to be shown to a group of operators who had little education, and were not knowledgeable about the design and manufacturing processes, a different content knowledge level would be needed. This knowledge level identifies an entry point for the content to be presented.

Another key question that you should ask during audience analysis is: "What is the viewer's need-to-know level?" By that we mean, how important is the viewer's need for the information being presented? This is a significant element during the scripting stage because it will help determine the influence of motivation. The script may have to have motivation factors built into the program to persuade the viewer that there are advantages to viewing and learning the material. (More on this in the scripting section.) Most important in defining the need-to-know level is the fact that it can help in the determination of just how substantially resources should be devoted to the training experience. The higher the need-to-know level, the less effort you have to devote to presenting the training. If the audience has a strong need for the information being presented, you do not have to present a sophisticated, expensively produced program. There is a place for the talking head! You have to measure the need-to-know level against the resources being devoted to the project.

An example of the importance of identifying the viewer's need-to-know level and how it affects production values can be summed up by a story that anyone who does much air travel can appreciate. As the airplane gained altitude, the pilot turned on the cabin speaker and started the usual Welcome Aboard speech." The passengers slumped down in their seats, tuning out the world for another boring flight, when just then the pilot said, "Oh, my God!" and the speaker fell silent.

What we have here is a very high need-to-know level. The passengers would not care whether the pilot came out with a film, videotape, slide/sound program or flipcharts. Their need-to-know level was very high. They were less concerned about how the message was to be delivered than what the message actually was. Their concern was information. All the fancy instructional design and media systems would not matter. Talking heads would do. Just give them information!

My point is: before you invest in expensive design, strategy and media, analyze the viewer's need-to-know level and weigh that against the resources to be invested. Quite often, because of the viewer's desire and need to receive the information, a very simple presentation may serve just as well as a sophisticated, expensive videotape production.

SETTING OBJECTIVES

Since you are trying to create a change with training, objectives should be written in such a way as to describe the desired performance, the conditions of the performance and the standards of the performance. Objectives should be described in terms that are measurable. And they need to be realistic—goals that are achievable by the target audience.

Program objectives, when well-written, become the guidelines on which all further program development will be based. They force you to think through clearly and state exactly what you are expecting your program to do. Objectives provide a clear statement of the results you are expecting the audience to accomplish. And they provide guidelines for videotape production decisions as you will see in Chapter 8. Writing clear instructional objectives is not an easy task, and it is not the intent of this book to teach you how. But I recommend *Preparing Instructional Objectives* by Robert Mager (Belmont, CA: Pitman Learning, Inc., 1975). It is a classic reference for instructional designers and provides excellent guidelines for the preparation of objectives.

DEVELOPING LEARNING STRATEGIES

The learning strategy or activity provides the vehicle for the training experience. The instructional designer tries to select the strategy most likely to achieve the desired result, whether it be lecture, group discussion, demonstration, role-playing, site visits, games, case studies or programmed instruction. It is generally accepted in the education community that there are three types of learning: (1) cognitive, involving facts, figures, concepts, principles and procedures; (2) psychomotor, involving motor skill development; and (3) affective, involving attitudes, values and motivation.[2] Each kind of learning requires a different strategy or activity. The strategy you select should be the most efficient way for the learner to achieve the desired learning objectives.

It is important not to confuse instructional strategy with instructional technology. The strategy is the method by which the learning will take place; the technology is the delivery system chosen to accomplish the strategy. A training event may require multiple strategies and technologies to accomplish its objectives.

Once the strategy has been selected, the learning objectives are sequenced to build on previous learning and to hold the interest of the learner. It is important always to keep the learner in mind when you are setting the objectives and developing the instructional strategy. The learner must have the resources to accomplish the objectives and must feel comfortable with the strategy chosen.

With the audience, objectives and instructional strategy defined, you must next select the delivery system that will become an efficient training method. The delivery system should involve the learner in the training experience and help cause the learning to take place. The system becomes the tool used to achieve the result, and in most training situations a single technology is ineffective. The designer must choose multiple systems to accomplish the objectives. In industrial training, for example, designers may choose multiple technologies, such as videotapes combined with workbooks, slides or transparencies; videotapes combined with demonstration and photographs; or slides combined with audiotape and workbooks.

MEDIA SELECTION

Media selection requires the designer to consider the relative values of instructional objectives, strategy, resources and budget. It is an imprecise process that often ends up being heavily influenced by media preference and budget. Ideally, the program objectives, audience and instructional strategy will influence the media selection significantly. One factor may outweigh others, but each should be analyzed fairly for effective media selection.

According to Ronald Anderson:

> The problem of media selection has been further complicated by a tendency for course developers to consider media selection as an isolated and independent function that is undertaken at some point well along in the instructional development process. This viewpoint has sometimes resulted from attempts to make the media selection process as scientific and exact as possible. Although the goal seems worthy, the present reality does not permit scientifically precise decisions.[3]

Scientific precision is, indeed, hard to achieve. Too often, attempts are made to draw up fancy charts and graphs for the selection process, but it is not that simple. The choice is really based on the resources and budgets available in a particular situation together with the objectives of the program and the characteristics of the audience. Therefore, knowing the advantages of each of the media allows you to weigh each against the other elements and make an intelligent decision. Anderson's book provides details of each of the media and helps the program developer narrow down and simplify the media choices.

Objectives and Strategy

The objectives become the cornerstone of the entire project. Each decision you make during program development and media selection must be influenced by the objectives. Therefore, the objectives of the program should be explicit and written in measurable terms as stated.

Clear objectives help answer some basic questions about the learning experience. What are you asking the students to achieve? How will you know if they have achieved it? What standards are you basing the performance on? The objectives really set the stage for the learning to be accomplished. They give direction to the learning and provide guidelines against which to measure the learning that has indeed been achieved.

The objectives and strategy are based on the type of learning that will take place. If the objectives are to teach facts, figures and procedures, then perhaps graphs, charts, numbers and word comparisons will be used. Attention to letter size and color will be important. This may require planning for graphics and design. With an emphasis on numbers and graphs, the material would lend itself to a medium that treats color, detail and numbers well, such as slides, because of their high resolution characteristics and excellent treatment of color.

In contrast, if the learning is primarily psychomotor, where importance is placed on motion and modeling techniques, video will work well because video can convey motion and has the capability to model proper performance. If the learning is to be primarily affective, video would again work well because it can evoke emotion and attitude through close-ups, action and music.

Each medium has distinct attributes that treat such elements as motion, color, pictures, sound and detail with varying capability. Matching the attributes with the learning requirements forms the basis of media selection. You should thoroughly analyze all delivery systems before you make a final selection. Each medium has certain advantages and can treat objectives in unique ways. Too often, because of personal preference or budget, a medium is selected first and objectives are developed around its strengths or weaknesses. A training program that is less than fully effective results.

Although ideally the medium should be chosen for its instructional effectiveness, it must also be cost-effective. It is important, therefore, to analyze each factor for its impact on the media choice.

Resources

Resources of staff, budget, equipment and time will all influence the media choice. Obviously, if the resources are not there or they are very limited, the choices for program delivery will be limited. As we learned in Chapter 3, it is costly to produce good video training programs and the resources of production talent have to be available, so this element dramatically affects the media choice. You have to weigh the resources required to produce the program against the results you are trying to achieve. A one-time training event to a limited audience may not justify creating a fully-scripted video program. But if that one-time event has the potential of attracting new investors or if it will create large sales figures, the event may justify a substantial investment of resources. The evaluation process becomes an individual choice based on individual circumstances. Each training event will be different and will have different objectives, so each selection process will become unique to the particular event.

Equipment, staff and facilities have to be addressed when considering resources for the media decision. Quite often a media decision is made on the basis of equipment available to the target audience. A network of video players may be in place throughout the organization. This network then creates an easy system to access for training program delivery and the media choice is ready-made. But if the equipment is not in place, this becomes a serious issue to face. How will the target audience receive and see the program? Some kind of guarantee must be built-in to ensure that the audience can access the program.

Is there talented staff available for program development and production? Obviously, it takes talent to create effective programs, especially video-based training programs. Staff resources have to be considered.

Facilities will play a key role in the media selection process. Does your organization have the production facilities available to produce media? Cost factors of production, equipment operation, maintenance and facilities upkeep will have to be figured into the decision process. If your organization does not have the facilities or staff to produce media, and it has been decided that a videotape is the best delivery method for the training program, outside resources will have to be contracted for production. If your organization does not produce many programs in a year, contracting with outside production resources may be the best way to go.

Video managers will agree that you should be producing at least 20 programs a year before you can honestly justify an in-house production facility.[4] By using outside production resources (for any medium) you have to budget only for actual programs produced, and the overhead of maintaining a facility is eliminated. Going outside for media production may cost a little more per program, but on a yearly budget this cost may be much less than supporting a production staff and facility. (Chapter 3 details operations costs for production and facility management.)

The element of time will play an important part in the decision process as well. The amount of time that becomes available to create a training event will often dictate the media choice. Video production does take time to design and execute. If fast turnaround is dictated, a lecture or a video conference may be in order. Again, each event will have its own limiting characteristics.

Resources, budget and deadlines all have to be weighed against the program objective. Surprisingly enough, budget is not always the single element that will influence the decision. If budget is available, time may not be. So each element will affect the route you take. You have to ask, What medium (given the time, budget and resources we have) will create the results I want in the most efficient and economical way?

Hardware and Software

As we discuss media and the choice factors, it is important to understand the differences between hardware and software. For our purposes in this book, hardware is the display or playback device, and software is the program material. The program itself, the slides, film or videotape that is actually carrying the message, is defined as the software. The playback equipment, projectors, TV sets and videocassette players are the hardware. Thus, you have to consider both hardware and software selection in your media choices. When you make a media choice for a particular training situation, you choose the medium as a whole, both hardware and software, so you immediately become involved in the aspects of what that medium represents. Distribution of hardware with all of its logistics, maintenance and scheduling responsibilities has to be planned for in the media selection process.

DEFINE TRAINING OR INFORMATION

One last consideration that should be remembered when developing programs is whether the intention is really to provide information to the viewer or to create a change in the viewer's behavior. Many problem tapes or programs that do not work well do not have a clear, defined purpose or goal. Training programs differ from information programs because each tries to achieve a different result. An information program is designed primarily to create a better understanding of a job or to bring an awareness to a subject. It may influence a better work environment or attempt to affect morale. Information is usually helpful to know, but it is not absolutely necessary or critical to job performance. After viewing informational programming, the audience will not necessarily be held responsible for the information.

An example of an information program is an employee news program. Viewers are not usually held accountable for any action after viewing the program. They are not held responsible for the content of the program. However, these programs can be justified through their communication objectives. News programs tend to be a great way to deliver information in a visually interesting manner. Television adds credibility to messages that might otherwise go unread in a memo. The successful approaches to news programs that I have seen tend not to take themselves too seriously. They stick to visually interesting stories and are delivered in the atmosphere of a staff meeting with a supervisor present who responds to and encourages discussion.

On the other hand, a training program has to generate accountability, and an action or result is expected after viewing the program. The viewer is held accountable for the content and is expected to perform differently on the job after viewing the program. The viewer is tested or audited for a specific change in performance, behavior or attitude. An old Chinese proverb helps me to define what good training should do, "Tell me and I'll forget. Show me and I may remember. Involve me and I'll understand." Effective video training involves the viewer and creates understanding.

You should determine whether the program will be training or information, because the end result or expectation will be different for each. Hence the designs of the two kinds of programs must differ.

For example, in information programs, such as corporate news shows, corporate messages, employee benefit and orientation programs, video producers try to create excitement and impact through music, fast pacing, glamorous talent and attractive visuals. The impact the message has on the viewer is quite often as important as the message itself. The person delivering the message is often as important as the message itself. The impact of the program often becomes the overall objective, the goal. And the evaluation of the informational program is based on the impact it has on the audience. This certainly is not wrong, but this objective has to be defined at the outset.

By contrast, training programs are designed to create specific results that will effect a change, behavior or attitude. You must carefully design the program so that the changes in your audience will occur. The programs need to be designed so that they are precise, to the point, using pictures and words that add to understanding. Visuals are carefully chosen and created to clarify and reinforce the message, not to interfere with the message but to add to it. You will become less concerned about creating a pretty picture and more concerned that the images are accurate, not misleading, and that they create a clearer understanding.

Training programs use music judiciously so that it does not distract from the learning. Talent is selected carefully for credibility and accuracy of delivery. Pacing is purposely slowed down in critical areas so that the viewer has a chance to digest the information and so that all viewers can follow along. Since you are expecting an action to take place after the viewing, you have to plan for different levels of acceptance by your audience. Each person viewing the program will receive it a little differently and at a different pace. This acceptance has to be planned for.

In training programs you are less concerned with entertaining than you are with teaching. Effective adult learning principles apply.

USING ADULT LEARNING THEORY

Adults react to teaching in varied ways, but most adults who respond positively to learning do so because solid learning principles were applied. It is important that producers of video training tapes for adult audiences follow the rules of adult learning theory.

Seven basic areas must be addressed in designing and producing successful video training programs. If you consider the following principles when programs are designed and produced, clearer communication will take place, and the producer will be closer to meeting the program objectives.

Motivation

In order to benefit from a training tape, viewers must feel that there is something in it for them. Understanding the audience's need-to-know level is an important design step on the road to creating a success. What's in it for the audience? What is the motivation to learn? Does motivation already exist or does it have to be written into the program?

Accurate identification of the motivation level will affect the learning and acceptance level of the viewer. Some audiences are highly motivated to learn and accept the information you present. Others appear to turn off to everything that is said. In many cases, an audience will display a mixture of acceptance and rejection. Understanding the motivation level helps you create a concept of delivery, or a way of packaging the program, that increases the program's chances of success.

Pace

Adults learn most effectively when they are allowed to learn at their own pace. Individuals need time to accept new ideas and weigh them against their personal experience. Well-planned pacing allows this process to take place. When designing and scripting linear programs, you should be sensitive to the pace at which you deliver the information. You need to allow for slower acceptance by some viewers while at the same time being careful not to bore fast learners. This is a difficult writing skill to master, but it is crucial to the program's acceptance.

Active Participation

When possible, the learner should be involved in designing the learning experience. Select a few participants from the target audience and include them in the design process. Not only will this help assure you that your program is on the right track, it is a good way to create an evaluation instrument. Your target audience can help you decide on what can be evaluated in the program.

Remember that when writing and producing training programs, you may find it easy to work in a vacuum, closing out others in the process. This can cause serious problems; work as a team to solve problems, and let the training tape be the product of the producer, client and audience team.

Programs that elicit the audience's active participation are more successful than those that do not. It is wise to have stopping points in the tape to allow for an exercise of one type or another to take place. Because linearity does not offer the flexibility of interactivity, linear programs must be designed carefully to bring the viewer into the learning event.

Conciseness

Respect the fact that the viewer is devoting valuable time to the training program. No matter how motivated they are, audiences prefer short programs that address the problem accurately and deliver the message efficiently. I believe in producing several short tapes that focus on aspects of a lengthy subject, rather than trying to accomplish too much in one long, boring tape. I support the 15-minute program and believe that a well-designed, well-executed program of this length can effectively deliver a lesson that would take an hour to deliver in a classroom.

Problem-Centering

Programs that are problem-centered rather than content-centered are received better by adult audiences. By designing training programs that take a problem-solving approach, as opposed to delivering straight content only, the producer stands a better chance of communicating effectively with the audience. Of course, not all video training programs will lend themselves to this approach, but where appropriate, present training in such a way that the audience is taken through a problem-solving experience. This ties in with the active learning theory. The audience is apt to get more involved if problems are presented and solutions generated.

Applications

Adults are eager to apply the new skills they have learned. Quite often this is difficult to control, especially for the video department, which often does not get deeply involved in program distribution. If control of the application is impossible, then time to apply the new ideas should be built into the training experience. This can be written into the program by allowing for such activities as problem-solving, interviewing or role-playing to take place as part of the training experience.

Climate

The proper learning climate must be created. It is important that the video department acquire information about how and where the tape will be viewed. If the program will be shown in a positive, controlled learning environment, less program time will have to be spent fighting unwanted factors, such as poor viewing conditions, office politics, unrealistic training expectations or time constraints. The producer must know the conditions of viewing in order to write a program that will teach the audience and meet its objectives.

INFORMATION	TRAINING
nice to know	should change performance, attitude or behavior
may improve working environment	
may affect morale	
viewers not accountable for a specific action	viewers accountable for a specific action

PRODUCTION TECHNIQUE	
high entertainment value	visuals selected for effectiveness
production values tend to become as important as the message	talent selected for credibility and accuracy
	effective adult learning theory applied
	pacing allows for message acceptance
talent often selected for impact or entertainment	production values
	carefully planned to create clearer understanding
evaluation often based on impact, not specific results	

EVALUATION

Designing training programs is a cyclic process that never really ends. Each step is evaluated and altered in a continuum that goes on throughout the development of the program. So the program is never complete. It is in a continuous state of change. The evaluation does not start when the program reaches the viewer. At that time it only offers further alteration options for the producer. I feel it is too late to ask if the materials are appropriate or if the presentation met its objectives after the viewer has seen the program. These questions arise during the design process.

Most important, however, is the need to have measurable objectives for the training experience, so that evaluation can take place. The evaluation process should relate directly to the objectives.

For technical training programs that require a training audit, observable actions (skills) should be evaluated as they relate to the objectives.

Sales techniques, behaviors, specific actions and processes that are taught in the video training program can be evaluated if the objectives are clear and measurable. The evaluation method will depend on the objective. But, as mentioned earlier, you should evaluate results before the program reaches the audience.

5

Scripting and Storyboarding—Phase 2

Scripting can be a difficult chore for those who do not do it full-time. But scripting, like most skills, improves with practice. In my experience with scripting, I do very little actual writing but a great deal of interviewing and outlining. My scripting method is useful for people who do not intend to become full-time scriptwriters but need to develop some familiarity with scriptwriting.

Writing is an individual, personal experience that cannot be learned from a book. But the mechanics of researching, outlining, visualization and script development are activities that you must master before you can fashion a functional shooting script.

Most of us in the field of training and video production have been hired for particular skills and not because of knowledge of a particular subject area or product. Scripting can be perplexing if the producer knows nothing about the content of the proposed program. It is your job to gather content, organize it logically, refine the information, visualize it to make it interesting to the eye, and in the end create a successful 15-minute program that will bring training results. In the scripting process, the general rule of thumb is that after you have gone through a sound design phase and a thorough scripting process, you will end up using only about 25% of your original, intended material. If you are doing your job correctly, asking the right questions and designing the program for the right audience, this 25% will be refined into a concise, highly-focused video presentation that makes the point and delivers the message with no extraneous material.

RESEARCH

The scripting process starts with research, and most of us are familiar with the common research techniques. Bill Van Nostran's excellent book *The Nonbroadcast Television Writer's Handbook* (White Plains, NY: Knowledge Industry Publications, 1983) has a chapter devoted to research methods. Van Nostran emphasizes that research will lead to answering the three-part question, "What do you want to say, to whom and for what purpose?" It is important to have a direction for research, and this question will give you a direction to follow. First, you need to gain a feeling for the total communication/training task; the initial question will allow you to focus this task. Second, it is important to know who the audience is. "The nature of the communication or training task changes radically based on the client's intended audience." "A significant portion of the research agenda may focus on gathering insight into the audience's level of knowledge or attitudes." And third, the answer to the important question of purpose "forms the foundation for delineating program objectives."

One of the best ways to do research is to interview individuals who are content experts or who will play an important role in program development. An effective technique for interviewing individuals is to record their responses on audiotape. The transcribed tape can actually become a working draft for the audio portion of the script, so you have almost immediately created words on paper. The interviewing process used for researching material leads to conversational information that usually would not surface in formal writing. This conversational style is important for creating a successful tape. As you learn about style and approach, you will see that to be received well by the audience a good tape should be delivered in a comfortable, conversational style. Remember that television is a one-on-one medium that gets the best results when the program appears to address the viewer in an informal, personal manner.

Another research technique is the on-site visit. This is particularly useful when the subject matter involves physical or mechanical content like a manufacturing process, a new product or a technical application. In these cases, seeing the process is vital. On-site visits usually include an interview on audiotape with the key people who will make the process work. It is also valuable to shoot photographs during the site visit. These photos aid in the storyboarding and visualization process. They can also assist in arranging for site logistics such as video equipment location, lighting requirements, staging and prop requirements. On-site visits are extremely important when more than one content expert or narrator will be recording the tape. If you plan to interview operators, salespeople or trainers on-site, it is best to record on audiotape the gist of the interview, so you can get a feeling for the personality of the individual being interviewed. This will help you develop a script that will reflect important characteristics of the interviewee accurately. On-site visits always reveal valuable insights, historical detail and personal perspective that sitting in an office writing a script just cannot do.

Quite often in the scripting process, you receive written material from the content expert or from the project requestor. This is usually in the form of an outline or formal description. If it is a detailed outline, a scriptwriter can then go back and ask more questions to fill in the gaps. Sometimes a writer may receive a written speech describing the process to be covered in the program. For some reason, people do not have such a fear of writing speeches as they do of writing scripts. When I ask a content person to write a script, he or she says, "No way. I'm not a scriptwriter. I can't write." But if I ask the scriptwriter to write a speech, the answer is, "Fine, give me a couple of weeks. I'll work on it." Actually, the two can be basically the same, and as long as you get the information that you need, either will do.

Another approach, if you have a content person who does not have the time to write a speech, script or detailed outline, is to have the individual recite content into an audiotape recorder. Talking doesn't seem like such a chore compared to sitting down in front of a blank page having to write descriptions. The audiotape will help create a conversational tone because the content person naturally talks to you on the audiotape recorder in a more relaxed, informal way.

The final suggestion, and probably the most difficult research technique, is the old library approach: going to the corporation library, the local library or into the manufacturing area, sitting down, and researching information out of books, catalogs or brochures. This approach usually does not generate the specific information you need, but can supply you with general background data.

During the research phase, you will also want to review what has already been written about the subject in the form of speeches, articles and studies, and what media already exist. I would guess that for virtually every training subject, a program has already been produced. The Association for Educational Communications and Technology offers a four-volume catalog of programs from the National Information Center for Educational Media. It contains over 60,000 titles. The National Video Clearinghouse, Inc., offers 40,000 titles and *The Video Source Book* contains over 40,000 titles. I have provided further program references at the end of this chapter.

Most programs that are made available for commercial distribution are designed to be generic, but they are generally well-produced and may meet the media need for the training event. Even if you do not wish to customize the program to your particular situation, just reviewing titles that are available can provide a valuable resource in giving you a perspective, style and approach for developing a new program that would meet your specific needs.

APPROACH

Once the research process has started, and as you become more familiar with the material, the content to be developed will start lending itself to a special approach for delivery to the audience. The approach will be the framework or theme that will deliver the content to the target audience successfully. Typical approaches would be a stand-up lecture with slides, a voice-over describing a demonstration or a new product, an interview or news show concept, a documentary, dramatic vignette, game show, plant tour, problem-solution approach, humorous scenes or the big talking head.

The audience is a determining factor in approach selection. You have to understand what theme the audience will accept. You have to determine what will work and will not work with the target audience. Will humor be accepted? Will an interview be credible with the audience? What format of delivery will work best in reaching the target audience? The script will grow from the theme, and the mood, music and talent will be developed based on the approach you select.

In seminars that I teach on designing and producing video training programs, I am continually asked, "What approaches work best for training?" It is always a difficult question to answer because the approach depends on a combination of elements, each influencing the decision. The audience is, however, one of the most important factors that will help determine the approach. But resources, budget and deadlines will also influence decision. It may not be practical to create a formal script where actors are hired to sell new product ideas. Or deadlines may limit how much time can be devoted to the project, and a talking head, therefore, might be appropriate.

Treatment

A treatment is a narrative description of how the program will appear on the screen. It will include the objective and an audience profile, but more important, it will describe how the information will be delivered—the approach. The treatment is usually a single page, and for some organizations it will become an actual proposal. The proposal will be presented to the client and upon approval, the project is set in motion. It is an important step in the process because approval on the treatment means that the objective, audience and approach have been accepted. This will provide direction and clearance for the project.

Content Development

The content you have gathered now has to be organized into a logical sequence. Training scripts can be organized three ways: in a time or chronological order, a topical order and a problem-solving approach. For example, with sales training you could organize the information into a problem-solving approach. The objective and concept of delivery will help in the selection of the organization of the material. Remember that with the television experience, the viewer will have only one chance to receive the message. The organization of the material becomes critical for successful reception.

As the outline is developed and content is defined for the target audience, the script will start to grow into a logical chain of events that will help your audience achieve the objective. But as the script grows it is important to start asking, "What exactly does the audience need in order to achieve the objective?" Content will grow and grow and the content expert, who is very close to the subject, will keep adding information that may not be necessary for the target audience. As writers, you must continually edit content and include only what is absolutely necessary. A training tape should be precise and to the point. You should respect the viewers' time and provide them with only the essential information. As pointed out earlier, videotape has the capability of condensing time. Through sharp writing and transitions you can present the required information in a very short time. Do this by honing down the information, and presenting through visuals and the spoken word only the information that is needed to reach the objective.

There are two schools of thought on how to proceed at this point. One is that you write the script first and then you visualize it. The second is that you visualize the information first and then write the script. I believe in the latter. After all, you have chosen video because of its visual impact, its ability to condense time and space, and its effective combination of the spoken word and pictures. Therefore, I do not believe in creating videotapes that are glorified audiotapes. Often visuals are placed into a script to support the audio, and if you turned off the video you would still receive the message. If this is the case, why not save money and just produce an audiotape?

If videotape is the chosen medium of delivery, use it to its fullest advantage. You should, therefore, think in pictures, pictures that will tell the story. In other words, think visually. Create a script that is strong visually and the pictures really do tell the story better. To this end the writer should address exactly what it is that must be seen to tell the story effectively. I approach this by jotting down visual ideas and picture sketches that I feel are needed to get the message communicated to the viewer. I actually create a storyboard or a sequence of pictures that depict the training event. Then I write words that support, clarify and enhance the visual experience. Of course, not all video training tapes lend themselves to this technique; but when you need a sophisticated training tape, think visually and let the pictures carry the message.

Whichever approach you use to create the script, remember to place yourself in the viewers' position and continually ask what they need both visually and narratively to receive the message and accomplish the objectives.

The script will be influenced by the design, audience and objective. The writer will use the approach to tie the content together and, through careful organization, will create a logical flow of information that the target audience can follow. The visuals chosen will accurately tell the training story in an interesting and lively way. And the words spoken will enhance the message, direct attention and help with transitions. Bringing together all of the individual elements of sound, pictures, transitions and music called for by the script creates the whole. And the training tape comes to life as a viewing experience that should motivate and teach.

For reference, a double-spaced, typewritten page will equal about one minute of programming depending upon how much visualization there is. So, if you are counting, you need to type about 20 double-spaced pages for a 20-minute program.

Page Format

A popular page format to use for video scripts breaks the page into two sections. One third of the page, usually on the left side, is devoted to visual information and is labeled "video." The other two-thirds of the page is devoted to the audio portion of the program. Label the program with a title or a series title, the client, writer, date, project number, etc. The date at the top of the page will change as the script goes through the approval process. (Fig. 5.1)

Type the words the narrator will read in upper and lower case, as if you were typing a business letter. This makes it easier for the narrator to read. All audio cues, however, such as music and sound effects, should be all upper case to separate it from the narration. This section will also be used to identify sound sources such as on- or off-camera talent, music mixes, sources of sound effects and direction of off-camera audio.

The video column contains descriptions of scenes, shots and angles. It will also contain directing cues for transitions, camera moves, lighting cues, graphics and all essential visual information that will be needed to carry out the video portion of the program.

Script Tips

- Write with emphasis on picture. After all, we are creating a video program not an audio program.
- Use words to support and enhance the visual message.
- Don't get wordy. Let the visuals carry the message.
- Use words that the audience is familiar with. Be prepared, depending on the viewers' familiarity with the content, to define words and provide an additional glossary.
- Use an informal, conversational style. Video is a personal, one-on-one medium. Direct narration to one individual in the audience.
- Use "you" and "we" whenever possible to establish an informal one-on-one experience.

VIDEO	AUDIO
	1 Such incredibly intricate, detailed work...
	2 and that involves so many different technologies?
	3
	4 (New music up and under.)
	5
QUICK OVERVIEW SHOTS OF	6 It's a fascinating process...one that requires
HYBRID PRODUCTION PROCESS	7 close coordination amongst scientists and
(TRACK PRODUCTION LINE).	8 engineers with backgrounds in electronics,
	9 metallurgy, ceramics, chemistry, physics and
	10 mechanics.
	11
GRAPHIC OVER B.G.	12 The creation of a hybrid product can be
	13 broken down into four steps: the design
	14 stage, circuit fabrication, component assembly
	15 and packaging.
	16
NARRATOR HOLDS UP PORTABLE	17 Before a hybrid can be designed, the engineer
TELEPHONE.	18 asks simply...what is the hybrid to do? What
	19 will be its function? Will it be used in the
	20 guidance system of a missile? Or will it
	21 be used in an automobile...or to make a home
	22 appliance smaller?

Fig. 5.1. A page from the script of a video training program entitled "Hybrid Technology—At Work Everywhere," which was produced by Burr-Brown Corp. for the International Society of Hybrid Microelectronics.

- Use plenty of pauses. Pauses are used for emphasis and for editing. And with pauses, the audience has a chance to digest the information.
- Write in transitions. Transitions carry the viewer from one event to the next and become an important element in creating successful tapes that flow naturally, without visual disruptions. Transitions add to the continuity of the message and should be considered a part of the scriptwriting process.
- Read the copy aloud. That will give you a feeling for timing, transitions, information flow, conversation style and believability. The audience will hear the script, not read it, so it has to be appealing to the ear.

Script Approval

The first attempt becomes the "rough" script. This should be sent to the client for approval. Usually, in place of a written script I send clients an audiotape of the script with a storyboard. The final product is going to be a videotape that the audience will see and hear. The audience is not going to read a script. They are going to listen to it, so your clients should experience the script in the medium it is intended for. Rough scripts in printed form are often edited by clients as if they were to be published in printed form. An audiotape will present the script as it is to be presented to the audience, and clients should evaluate its effectiveness as they hear it.

Changes are a common occurrence in scriptwriting, so be prepared to face many changes at this point. With a personal computer and a good word-processing program, the scriptwriting phase can be streamlined. Some software products on the market are specifically designed for scriptwriting. The "Power Script" program by Comprehensive Video is an excellent example.

Storyboard

The storyboard is a sequence of simply drawn pictures that visually represent the program. An important element, it allows the producer to share visual choices and creative approaches with the client, get a feeling for the pacing and timing of the program, and "edit" the program for visual continuity and clarity of message. It also allows producers to "direct" the program on paper for overall visual and oral effectiveness.

The advantages of Hybrid Microcircuits are:
The short design time.......
production costs are low.

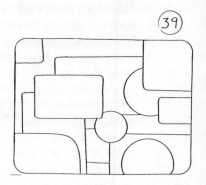

Why? Because Hybrids by their very
nature......
changing needs of Microelectronic
Technology.

Because of these and other advantages,
Hybrid Microcircuits.......
have such a predictably bright future.

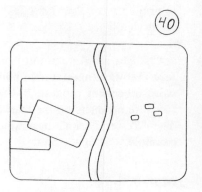

The driving force in Microelectronics
today.......
greater reliability and cost reduction.

Fig. 5.2. A portion of the storyboard for a video training program entitled "Hybrid Technology—At Work Everywhere," which was produced by Burr-Brown Corp. for the International Society of Hybrid Microelectronics.

The producer, often enlisting the talents of an artist, depicts through simple drawings the script as it will appear visually on the screen, frame by frame. The audio portion of the program will be typed in under each picture. The storyboard will have a picture of each significant visual event that fairly represents the script. These events include video locations, shots, angles, graphics and transitions. The storyboard visualizes the completed program on paper. For the first time the clients "see" the visual images that tell the story. Graphics and effects can be presented to be checked for placement, clarity and effectiveness. Location, props, scenery and talent placement can all be checked before final decisions are made. (Figs. 5.2, 5.3)

Some approaches, such as the "talking head" or interview programs, will not require a storyboard, but more creative programs usually require at very best a simple storyboard. It is best to communicate visual ideas on paper. Paper is cheaper than tape.

Fig. 5.3. An artist prepares a storyboard for graphics, each of which will be sketched in advance so accuracy and effectiveness can be checked.

6

Pre-Production—Phase 3

After the script and storyboard have been written and developed, the next phase of producing video training programs is the pre-production phase. In this phase the shooting script, production requirements, talent arrangements, shooting schedule, graphics, site survey, set, props and training material are all developed.

SHOOTING SCRIPT

A complete shooting script is divided into scenes and shots. Quite often scenes are determined by location, major sequences, camera positions or learning points. Each script lends itself to some logical way of breaking up into smaller, identifiable sections.

If a new product introduction were to be videotaped, for example, the script might be broken up into scenes according to three major locations called for in the script. These might be exterior shots of the manufacturing plant, the interior of the manufacturing plant and a classroom.

The script sequence of our example product intro could be:

1. exterior of plant for opening
2. classroom where host will present product demonstration
3. manufacturing area where product is made
4. exterior shots for closing and credits

The four major sequences or events, according to location, would become scenes one, two, three and four. Each scene would then be broken up into shots according to camera positions. Camera shots are usually determined by the content and are close-up, medium and wide camera angles.

In this example the shots required in the script could be:

1. wide shot (ws) exterior of plant
2. close-up (cu) company sign
3. wide shot in classroom
4. medium shot (ms) host holding product
5. close-up product
6. medium shot host
7. wide shot manufacturing area
8. medium shot host closing program
9. exterior wide shot for credits

In order to produce this program efficiently, the scenes and shots will be recorded out of sequence. Recording out of sequence allows the producer to be efficient with camera set-ups, equipment movement, talent and environmental conditions.

In this example it would not make sense to haul the equipment outside for the opening exterior shot, move back inside for the interiors and then go back outside for the closing credits. While outside it would be best to shoot both the opening and the closing, even though that is not the sequence of events according to the script.

Similarly, in the classroom you would light for the wide shot first. The wide shot establishes the mood or "look," and the lighting exposure for the room and thus for all of the room shots. The wide shot becomes our "establishing" shot. All other shots would then be referenced according to the lighting in the wide shot.

After shooting all the shots in the classroom that require the wide angle, whether or not they are in sequence according to the script, you would then move the lights and the camera in and shoot all of the medium shots because this shot would require different lighting and camera positions. Finally, when the medium shots are complete, all of the close-ups are lit and shot. By shooting out of sequence you are saving yourselves a great deal of work moving the equipment back and forth.

Your sequence for shooting the example program would be:

> Scene 1, shot 1 = ws exterior
> shot 2 = ws exterior for credits
> shot 3 = cu company sign
> Scene 2, shot 1 = ws host
> shot 2 = ms host/product
> shot 3 = ms host
> shot 4 = ms host/closing
> Scene 3, shot 1 = cu product
> Scene 4, shot 1 = ws manufacturing

This is called "shooting out of sequence." Later on, all of these scenes and shots will be edited together in correct sequence according to the script.

The shooting script will also have major camera moves, audio cues, props and graphic notes listed. A smart practice is to retype or "cut and paste" each shot or scene onto one page.

After the script has been broken down into scenes and shots, it will be put into a logical sequence for shooting. Your program may require that the manufacturing shot be done late in the evening when there is a light working crew and your equipment will not cause too much disturbance. The exteriors should be shot in the late afternoon when the sun casts a golden glow onto the sign, and the scenes with the host should be shot in the morning when he or she is fresh. These required shooting times will create our shooting schedule. The shooting script is then arranged into shooting sequence according to our time requirements. If you were shooting your whole program in one day, you would shoot it in the following sequence: scenes 2, 3, 1 and 4.

Shooting out of sequence presents interesting problems of continuity, especially if you are using nonprofessional talent. It is difficult to grasp the idea of shooting out of sequence, so your host will have to be coached properly. The director will have to keep track of what is said in each shot to be sure there is continuity for the editing process. Talent movements, product placement, clocks on the wall, pens and pencils, etc., all have to be noted and kept track of to assure the continuity and flow of the script.

After the shooting script has been developed, your next step is to time each shot. This is essential in order to determine the amount of video to be shot on the scene. This is especially true if the script is mostly voice-over narration. By timing each shot, you will know how much footage to shoot in the field. If the length of shots is not noted, we may come up short of footage needed in the editing process to cover a narrated description of action. It usually takes longer to describe an action than it takes to do the action. Thus you have to have enough video to cover the narrated description of the action.

To be sure that enough footage will be shot during the videotape session, the audio for each shot should be read aloud and timed. It is also smart to shoot "inserts" or "cut-aways" while the lights and camera are ready. Insert shots are "cover" shots used in the editing process. They are close-ups of the talent nodding, smiling, looking to the right and left, up and down. Or they are close-ups of hands and props that are not really called for in the script, but that might be needed in editing to cover an awkward edit, camera change or script change. As production pro Lon McQuillin so wisely states about shooting enough material for the editing process, "You can't shoot too much material; you can shoot too little."

PRODUCTION REQUIREMENTS

Production requirements, such as location, talent, equipment, crew, sets, props and graphics, are determined by the script.

Location

The location is set by the requirements of the script. But the location can in turn influence the script and can actually create or enhance the story line.

When the location has been selected and agreed upon by everyone involved, your next step is to make a site survey. This consists of actually visiting each location called for by the script in order to determine lighting, sound and logistics requirements. If special lights will be required, the site survey should include the location of all power outlets and breaker boxes. It is critical to locate circuits and determine how much lighting can be operated on each circuit. The script will dictate the lighting effect needed to get across the intended mood or emotional impact and enough power will have to be obtained to create that effect. The keys to the circuit breaker box must be obtained because a circuit will most likely be blown during the shooting. I always feel that if I don't blow at least one circuit during a shoot, I'm not doing my job.

The site survey should also determine the sound requirements. Special sound baffling may have to be used if there are air-conditioner vents or doorways. The trained ear will also be alert for unwanted outside noises, loudspeakers, traffic noise, fluorescent light fixture noise and nearby mechanical noise that will interfere with the sound recording.

Camera positions should be determined and problems can be dealt with, such as windows or doors in the shot, or busy or distracting backgrounds. These problems should be resolved so the production will not be slowed down on the day of the shoot.

The site survey may also suggest special equipment requirements, such as power generators, dollies, radio microphones or reflectors. And it should establish the logistics of the shoot. These might include:

- transportation to and from the location for crew, talent and equipment
- equipment transportation around the location
- equipment security
- location of props and demonstration equipment
- clearance for location use
- meal arrangements
- bathrooms and dressing rooms
- telephone
- power requirements and locations
- house engineer support
- measurements for cable runs

Once the site survey has been made, the script may have to be modified to reflect the logistics involved. Some scenes may need to be shortened and others lengthened because of problems that may be encountered at the site. The shooting schedule and the script may change as well because of unforeseen problems revolving around a certain location. The site survey is important and should be done as early in the pre-production phase as possible. (Fig. 6.1)

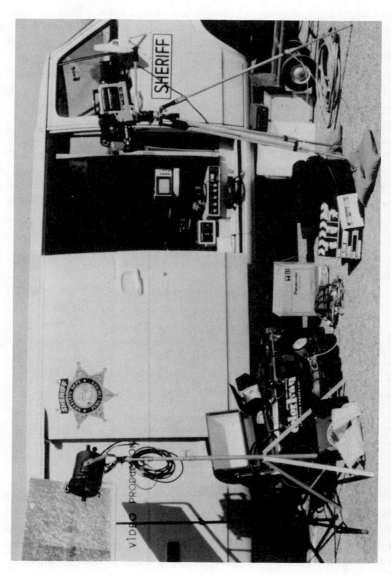

Fig. 6.1. The transportation of all the equipment required for a remote production must not be overlooked during the equipment planning process. (Photo courtesy of Pima County Sheriff's Dept., Tucson, AZ.)

TALENT ARRANGEMENTS

The script will call for using either in-house, nonprofessional talent or professional talent. In either case, it is important to send the script to the chosen talent in advance of the shoot. Quite often for training programs, the talent is the content person who helped develop the script, and who therefore will be familiar with the script's requirements. Even such a person may still have to be briefed on what to wear (more on this in the production design chapter), where to appear for the shoot and how the production schedule has been set up.

Working with nonprofessional talent is an art in itself. Two keys to success in this area are patience and communication. In the excitement and stress of production it is easy to get upset with nonprofessional talent. But patience will go a long way toward avoiding flareups and personality conflicts. It is amazing but true that a marketing manager, product manager or company president can become a seasoned director overnight. Patience along with clear communications is the key here.

Nonprofessionals are out of their element standing in front of a camera. The workings of videotape production are foreign. They have no idea why the tape was stopped, a light moved or a camera angle changed. Good communications with the talent on what is going on at all times will help alleviate any fears and will calm the nerves of a shaky, first-time talent. Let the talent know what is going on, that the change was not their fault, and that all the efforts with the details of lighting, audio and staging will make them look their best.

When it is decided to use professional talent, a casting session will be scheduled, parts of the script read (ideally in front of a camera) and the appropriate talent chosen. A contract will be drawn up, detailing the time required for the production of on-camera and narration parts, location details, costume required and costs involved. The script is often sent so that the talent can start preparing for the shoot.

EQUIPMENT

Script, location, budget and internal resources will dictate equipment. Each project will have different equipment requirements, just as each facility will have different resources and budgets, and these must be decided upon and planned during the pre-production phase. Some of the requirements might be reflectors, power supplies, extra lights and grip equipment, shot-gun microphones to cover wide-angle shots, camera dollies, filters, extra battery packs and monitoring equipment. All of the special equipment requirements have to be planned, scheduled and procured in advance of the production. (Fig. 6.2)

The decision for the recording format becomes a complex matrix of budget, internal resources, script and distribution requirements.

Sometimes the choice is already made by the type of equipment available at the facility. Other times the format selection is dictated by budget or program distribution requirements. But time has to be planned into the pre-production phase to make arrangements for procuring the equipment needed.

CREW

Like everything else, the crew requirements are dictated by the script, budget and internal resources. Whatever the crew requirements end up being, the arrangements of scheduling and contracts take place during the pre-production phase.

An important aspect of crew arrangements is communications. Pre-production meetings that explain the desired outcome of the program, equipment involved, location requirements, talent considerations, schedules and deadlines, limitations of the production and the agreed-upon "look" of the program should be communicated to all involved. My experience has proved to me that it doesn't pay for the producer/ director to work in a vacuum. All members of the production team can make creative suggestions and important contributions that are easily overlooked by one individual. The production of a training program is truly a team effort. Everyone involved contributes to the end result. To take full advantage of the team, clear communications must take place. Make special efforts to have plenty of production meetings and create a climate for contribution. This effort will pay off, resulting in a successful program.

During a typical production of a training tape produced in-house the crew consists of the producer/director who usually does the camera work, a technician who operates the recorder and helps with the lights, and where appropriate, a production assistant who helps the director with all of the pre-production arrangements, watches continuity during the shoot and assists in the post-production phase. To some readers this typical three-person crew will seem extravagant; to others it would seem impossible to produce anything with so few people. But most

Fig. 6.2. Acquiring the proper equipment for a video production is essential. Here, during the taping of an Adolph Coors Co. production, a microphone boom was required to capture the talent's voice.

organizations producing video for training operate very efficiently and effectively with this size crew, producing some outstanding results.

SETS AND PROPS

If sets are required, this is the time to begin design and construction. Even the simplest set requires time to build, set up and light. Often the room arrangement where the tape will be shot will be left to the last minute. This will cause problems and waste a great deal of production time. If a classroom is to be used for the production, draw a sketch of the proposed room arrangement, and think about camera and equipment placement.

Once the set or location has been selected, lighting diagrams should be drawn for the studio or for the location production. Plan lighting instrument placement. Design the lighting "look" well in advance so that power and special equipment requirements can be arranged.

A word about props. Props tend to become a real problem during production. While they seem to be an insignificant detail, props will get you. A detailed list of all the props that are needed for the entire production should be created. You should make this list far in advance of the production, not the day before. The prop list will grow. What at first seems to be a simple list of a few products will often become a large list of problems.

Loaner props are often used; antiques, older products, historical photos, product parts, household items, special desks, books, lamps and product application examples all take time to arrange for. Similarly, you will need time to make phone calls, make pick-ups and deliveries, and arrange for purchase requisitions and equipment contracts.

Be sure to get prior approval for the use of all products. The day of the shoot is not the time to check if it is all right to use a new product. This approval process is lengthy, so plan well in advance for the acquisition of all props.

GRAPHICS AND SPECIAL EFFECTS

Special effects and graphics should be planned far in advance of the production also. All of the elements of slides, charts and graphics should be produced before the shooting begins so that after the field or studio production takes place, all graphic elements can be shot and put on tape.

The graphic elements need to be designed and planned. The artwork for slides, art cards and charts will require time to produce. The "look" of the graphics needs to be consistent with the desired "look" of the program and support materials. The instructor's guides and student manuals need to be created with a coherent type style, page layout, cover design and graphic style. This image should be consistent with all graphics used for the training program. And the graphics used in the tape should match the graphics used in the printed support materials. Coordination of printed materials and video graphics can create an attractive, coordinated package that has a consistent image. The effort that goes into well-planned graphics, type style, photographs, page layout and illustrations will all contribute to a coherent training experience.

I believe strongly in creating a cohesive image that ties all of the training materials together. Experience has shown that a strong graphic design carried through to all training materials will help create continuity of learning, ease of reading and comprehension of materials, and ease of locating desired information. A complete training package works to create an effective learning environment and experience. It reveals a dedicated commitment to the training effort and adds to the viewers' attraction to the idea of getting involved with the training being presented. Graphics coordinated with the "look" of workbooks, brochures, exercises, examples, certification tests, promotion and videotapes work to create success. (Fig. 6.3)

If special effects are to be used in the program, they should be designed and visualized on the storyboard. Special effects could include: split screens, dissolves, wipe-on-wipe-off pictures, chromakeys and digital effects of picture zooms, and rotations and spins. These effects are usually added during the the post-production process; however, they should be planned in advance of the shoot. When in production, the director should know if the scene or shot will be used in an effects event during post-production, so that the shot can be set up properly. (More on this in the Production Design chapter). Special effects have to be planned in advance of production. Time during post-production will be saved if the effects are thought-out and planned. A well-designed storyboard will depict the effects events, showing image positioning, graphic placement and time elapse of all events. The pre-production phase is the time to plan for continuity of background colors for graphics and products,

Fig. 6.3. An artist prepares the artwork needed for a video training tape in advance of production.

so special effects will add to, not detract from, the message. Plan as far in advance as possible what the effects should look like, and how to get into and out of scenes by using graphics or effects. In other words, the entire program from start to finish—all details of effects, graphics, background colors, movement and event times—should be conceived thoroughly. Adding special effects can get expensive, in large part because of the expense in the time needed in post-production to create the effect. You will save money in post by being prepared; have the effect well-thought-out.

The importance of pre-production planning cannot be stressed enough. As a producer/director, you should devote your energy during production to working with the talent and capturing the image you want on tape. Energy should not be wasted on such minute details as finding an extension cord or a fuse box. All the details of logistics, equipment and location should be worked out in advance of production so that they do not become a burden on the day of the shoot. It is too late to worry about having the right connection or cable on the day of the shoot.

Tom Kennedy, an independent producer/director working in San Francisco, sums up the importance of pre-production planning this way:

> In directing a training program, preparation is every bit as important as in dramatic video programs. In drama, having all production details organized lets me work with an idea or concept—while in documentary and training programs, walking in prepared lets me become the problem solver who's often needed—as when an assembly line procedure turns out to be entirely different in practice from what it is in the company's manual that was used for scripting.
>
> In any program's production, the feeling of assurance that I've done my homework lets me focus on shaping the talent's performance. If I'm clear in what I want, my timetable, camera placement and end points for each shot, I can give complete attention to crew and talent as people and get the best performance out of each. The creative opportunities are realized when you're prepared to take advantage of them.

7

Production Design

In the design and production of video training programs the actual recording of the tape often receives the least amount of planning. We think of video as lights, camera, action! Much energy is put into the action part, but perhaps not enough energy is placed on planning it. Much emphasis is placed on turning on the camera and recording pictures that we hope will tell the story. But as we turn that camera on, have we actually thought through what we are trying to do with that picture? The script leads us down the path of the story line, but what about the actual pictures that we are creating? Are the camera angles, talent movement, lighting, sound and graphics that we are recording actually effective? Are they adding to or detracting from the message?

Video is a visual medium. We consciously or unconsciously stress the picture. But video really is a combination of production values that all come together to form the program. These elements are light, sound, composition, camera angles, movement, music, pace, special effects, graphics, color and talent. All of these affect the message. The video producer tries to capture them all on tape to create a pleasing program. But for what purpose? Why are all these elements coming together? Are they coming together for the sake of creating video, or are they all being orchestrated to deliver a message? This is where production design comes in.

Production values can be compelling; they may even interfere with the message. In today's world of sophisticated video electronics, are we starting to produce video for video's sake? If I notice the lighting, camera angles, sound effects or music, perhaps they are wrong. After all, training programs are not about the pretty lighting technique that creates a mood or the special effect that highlights a graphic. You must understand the power of video, and its overall impact on the message and the audience. You must use this impact so that it works for us and not against us. In video training, the purpose of the program (with all its production values) is to cause a change in the viewers, a change that may affect their jobs, their work habits or even their lives. The choices that affect the video production, therefore, become extremely important decisions.

Remember that the production values must help form the message. They add interpretation and help create the outcome of the program. Their impact on viewers is such that the message is or is not clearly received. And they definitely affect the learning that is to take place.

Lighting

Lighting for video is used primarily to illuminate the subject so that the camera can "see" it and pick it up on the pick-up tube.[1] It is used to enhance either a simulated environment in a studio or an actual environment on location. Light helps in composition, creating shadow, color, texture and form. Lighting can also be used to create a mood, a psychological condition that would not naturally be present. The video director uses light to create images that the camera will record. These images become the message.

Television is a flat, two-dimensional medium made up of only height and width. Lighting, staging (the correct placement of objects and talent), color and sound can create the illusion of depth. Lighting creates this illusion by separating the subject from the background, and by giving an impression of roundness and texture. It is important to control the lighting so that it does not interfere with the intended message of the training program and cause confusion. The eye is attracted to the brightest part of the screen, so you want to highlight the most important subject with light and make it stand out from the background or other images on the screen. Your subject should be the best lit or brightest spot. In training programs you should be less interested in creating a mood than in presenting your message in a clear, easy-to-see manner. In lighting training programs, attention should be paid to what you are emphasizing with the light you give the camera. You can control the eye on the screen by using light to emphasize and de-emphasize subject matter.

The object here is to create an image that "reads" well, that is not confusing or hard to see. The product being demonstrated should be lit so that the viewer can see it easily and the lighting does not become a distraction. Good lighting separation techniques can make the subject the center of attention so that the message is clear and the intended result is achieved. Reflections, bright windows, deep shadows and minimum separation can interfere with the intended message. The viewer should receive the images that make up the message effortlessly. It should be clear what is important and what is not important. The camera is not as discriminating as the eye. It will essentially record what it is given. It will not separate important speakers from less important speakers, which our eyes and brain automatically do. Confusing images cause the viewer to work harder to receive the message. Lighting becomes a tool for the video director to use in the effort to create a training event for viewers. (Fig. 7.1)

Sound

Sound is another tool that is used to create a video event. Sound can become a confusing, distracting, disorienting element in the program. It must be tightly controlled to add realism, depth and interest to the program, as well as credibility and emphasis. Good sound recording techniques should separate the voice of the main speaker from the natural background noise. It should highlight the voice track and de-emphasize unwanted noise. Microphone placement, sound control and mixing add a dimension to the program that will create believability. Learning points can be emphasized with sound effects. Recognition of machine problems in a maintenance program can be achieved by sound. Safety tapes can be more effective through good use of natural environmental sounds. And the creative blend of room ambience and narration can add to the believability of a location production. Recorded sound should be natural and clear, and it should enhance the message.

Composition, Camera Angles and Movement

Composition, camera angles and movement are so subtle that they are rarely perceived by viewers. But each of these elements adds to the overall effect of the program and to the message that is being created.

The television screen is really very small, so special attention should be paid to composition, that is, where you position subjects within that screen. The eye naturally moves from center screen to upper right, so

Fig. 7.1. During production, make up is applied to cut down on possible glare from the lights. (Photo courtesy of Mattingly Productions.)

the important subject should be emphasized by being in the center of the screen. It is awkward to read titles that are placed in the lower portion of the screen. These key locations tell us where to place visuals and subjects to emphasize or to de-emphasize them.

Another aspect of composition is concerned with where objects are placed within the scene. Are backgrounds, such as trees, wallpaper, pictures, telephone poles or windows, interfering with the foreground subject? Are backgrounds too busy so that they make viewers hunt for the important subject? Keep backgrounds plain and de-emphasized if possible.

Camera angles can also add to the message or create another unintended message. Especially working with talent, the angle a director chooses can distract viewers or interrupt the message flow. A camera angle slightly above eye level, which causes the talent to look up, can imply inferiority to the viewer. Likewise, an angle below the talent eye level can imply a dominant posture to the audience. Camera angles must be carefully planned and chosen to portray the subject best under positive light and create the best viewing angle for a demonstration to be seen.

Movement, whether of the camera or talent, also sends subtle messages to the viewer. Movement of the camera can emphasize a subject or de-emphasize a background. It can emphasize what the talent is saying by zooming in slowly, or it can de-emphasize a scene to make way for a scene change by zooming out. The talent can move toward the camera for emphasis or to highlight a learning point. Just having the talent walk around within a scene can create interest and break up the shot enough so that viewers pay more attention to what is being said.

With camera angles, composition and lighting you can control the background so that the foreground subject appears to be highlighted and to be the center of attention. This is important in training programs, where the subject has to be very clear. Again, the camera will not separate important material from unimportant material. You have to do this through composition, angles, movement and lighting.

Music

It is important to use the emotional qualities of music to their full advantage. Let them work for you and not against you. If you want to establish a certain mood, music can help create it. It can help you emphasize important learning points, graphics and action. Music can make the training program come alive. It can draw attention to what is happening on the screen and emphasize words, phrases or action. Music helps with transitions (moving from one learning point to another or from one scene to another). A little traveling music please! Music can help separate an important point from the rest of the program. And music can always get the attention of viewers at the beginning of the program or create continuity for a series.

Music can create problems as well. A tune that has pleasant associations for you may not have such pleasant associations for someone else. When you are moving into an important learning point, a recognizable tune may be a reminder for someone of an unpleasant experience from childhood instead of a preparation for the delights of a future accomplishment. Music can interfere with the message. Be highly selective, using indistinguishable tunes whenever possible. Use music to emphasize, not to call back memories.

Then there is the copyright issue. This may be complicated and costly, so you should use only music that you hold the rights to. These rights can be purchased through music libraries. Some good ones are Omnimusic, Port Washington, NY; DeWolfe Music, New York, NY; and Soper Sound, Palo Alto, CA.

Pace

Unlike interactive programming where the viewer controls the pace, in linear video programs the director controls the pacing or speed at which the viewer receives the information, according to Steve Floyd, author of *The Handbook of Interactive Video.*[2]

It quite often seems much slower than the pacing used for broadcast entertainment programs for a sound reason. The director of the training program has to build in pauses for the viewer to digest the information and relate it to past experience or to on-the-job experience (see Chapter 4). The director controls this pace and has to allow for different audience comprehension levels. Time has to be built in for viewers to read graphics, and for trainers to emphasize learning points and stress concepts. When a demonstration or explanation is presented, the director must know what the audience needs next and when to ensure comprehension by repeating or varying the presentation.

The director must understand the audience well enough to know how fast to present information. If the training objective is to teach a skill, the program will have to be paced at a rate that permits the whole audience to receive and comprehend the message well enough to learn the skill.

The director has to be able to anticipate the viewers' needs. A specific action or result is expected of the audience after viewing the program. The audience's grasp of what is being presented and a pace that is understandable must be assured. In linear video, viewers do not have the chance to go back and reread a paragraph or repeat a segment of tape. Usually video programs run from start to finish so pauses for emphasis, good separation of learning points and important action must be paced well for the learners to follow. Techniques such as using music to highlight concepts are desirable, and slow dissolves to slow down the pace are acceptable in training. It would be too much to ask of our audience to hang onto every word spoken during the program. At best key words, concepts and actions will be retained. Pacing helps emphasize these elements and allows time for retention.

Training programs should have peaks and valleys in pacing. Important action should take place at a normal speed to retain the audience's attention, but a slightly slower pace should follow for them to digest the important action. This action becomes a learning point, so it has to be separated from other action in the program. It is inadvisable to follow one important action with another and another in quick succession. Give the audience a chance to breathe, take in the information and relate it to their jobs. Be aware of the pacing you are creating, because this production value can add a great deal to the success of the training program.

Special Effects

Special effects such as dissolves, split screens, freeze frame, slow motion, character generation and computer animation can all be very effective devices in video production, but like any other production technique, effects can be overused easily and even interfere with the message. Some programs become so wrapped up in the effects that they forget the original objective of the program. Effects that are done for the sake of effect really do not add anything to the program. They can become so expensive that it is difficult to justify their use. Spins, whirls and animation are fun to watch, but in training programs, you have to be sure that they do not become so overpowering that they detract from the message.

Special effects can, of course, be valuable for some video training programs. If well planned, an animation event can contribute a great deal to the learning experience. It can reveal the inner workings of a machine, for example, and thus clarify the method of operation for the operators and provide them with insight into how the machine actually makes the product.

Similarly, the wafer fabrication training programs produced by Burr-Brown Corporation often use simple animation to reveal how a certain building process is actually forming the circuits on the semiconductor chip. This phenomenon is impossible to see with the naked eye, but through video animation it can be seen and understood.

The passage of time can be indicated with dissolves, or a tiny circuit can be blown up to full screen with an animated zoom. Graphic zooms, emphasized colors, glowing letters, moving parts, still frames of action, slow motion of action, dissolves and split screens can all contribute significantly to the teaching process if they are planned carefully.

Effects are normally created in the post-production process, but you really should plan for effects in the storyboard stage. You should think through each effect so that the scene can be shot in preparation for the effects to be created during post-production. For example, if a dissolve is to be used as a transition effect, the scene before the dissolve as well as the scene you are dissolving into have to be prepared with plenty of pre-roll at the front and the back of the scene. If more sophisticated effects are to be used, a detailed storyboard is a must so that each shot can be planned properly. But it is most important during the production design stage to understand fully why the effect is being used and how it will actually add to the program objective. "What a dissolve means— if in fact it means anything—is determined by context."[3]

Graphics

Some important points to consider when using color graphics for training programs are as follows:

- Type. Since the television screen is actually very small, the style and size of type are important for legibility. Large type sizes with simple (sans-sarif) style work well for television. Avoid fancy, condensed or expanded letters.
- Slides. Be careful when transferring slides of charts and graphs to video. Try to keep the slide densities the same. And remember, video does not have as high a contrast ratio as film. This can cause problems in the transfers. Keep to the horizontal, 3 × 4

television aspect ratio, and keep slides and graphs simple, not too busy. What appears legible on a slide may not look legible on video. Textbook graphs and charts rarely reproduce well on video.

Be sure that the visuals (word or graph slides) do not conflict with what is being said by the narrator. Let the visual carry the message. Do not repeat every word listed on the screen or overexplain the visual. And, finally, allow the visual to be on the screen long enough for the audience to read it. (Rule of thumb is to read it out loud twice).

Color

Certain colors, or color groups, seem to influence our perception and emotions in fairly specific ways. Although we still do not know exactly why, some colors seem warmer than others, some closer or farther away. Some colors seem to excite us, others to calm us down. Color influences our judgment of temperature, space, time and weight most strikingly.[4]

Subtle color distinction is not always possible in video as it is in film. Remember that color choices are important because colors create different responses. Reds, oranges and yellows appear to be closer to the viewer, and blues and greens appear to be distant. When combined properly in graphics, colors can create depth and pleasing contrasts.

Today's video cameras generally handle color well, but you should follow a few guidelines in dealing with color for graphs and slides. Red can cause smearing problems on lower-cost cameras. Try to use bright contrasting colors for charts, but be careful of colors that are too bright because they can call attention to themselves rather than to the information being presented.

Watch what you are saying in the colors you choose. This also applies to the colors worn by the talent. Costume patterns can become too busy, so watch out for the herringbone effect generated by small plaids and stripes on video.

Not all colors transfer the same way. The nature of film and
video is better suited to colors that translate into the middle
shades of gray. When shooting slides that will be transferred,
you achieve a better end product if lighting ratios are
controlled. Light as if you were shooting live video: lower
contrast and more light.[5]

Talent

In most training programs produced by organizations, nonprofessional
talent is used to demonstrate products or processes or to explain new
techniques. Quite often training tapes contain both nonprofessional talent
to demonstrate a skill or new product and professional talent to record
the narration. This is because familiarity with both the product and
process is important for the visual demonstration while a pleasant and
authoritative voice is important in reading the script. But you will soon
find that working with nonprofessional talent can be very difficult. The
time it takes to work with in-house staff to demonstrate a product on
camera or to deliver a few lines can often be reduced by using a
professional who is used to the television production process, can learn
quickly and can be depended on to deliver lines believably. The
justification for using outside talent should be looked at seriously and
not just thrown out as an unwanted expense or frill.

When working with nonprofessional talent, consider the following:

- Credibility. Does the talent you selected have credibility with
  the target audience? A person who knows a great deal about
  the new product may nevertheless have zero credibility with the
  audience. This is guaranteed to slow down the learning process
  and will influence the reception by the audience.
- Is the person delivering the message as important as the message
  itself? Quite often a person selected will enhance the program
  a great deal just by his or her appearance on tape.
- Patience and good communications are keys to working with
  nonprofessional talent (see Chapter 6). Corporate staff appearing
  in front of the camera are out of their element. Patience with
  their nervousness and lack of familiarity with the video production
  process will create a better working relationship. Your job is

to make them look their best on tape. Good communication with them and the crew is important in accomplishing this goal.

- Appearance. Nonprofessional talent may not realize how much their appearance may influence their message. It is important to coach first-time talent on what to wear for the production. Conservatism in dress is a good general rule. Plaids, thin strips and herringbone fabrics can cause moire patterns on television. Solid colored sport jackets and suits worn with plain, contrasting shirts and ties are smart choices. Avoid white shirts or blouses: they may cause lighting problems. Too much contrast can "wash-out" the face. Women should be careful not to wear too much jewelry as this may cause reflections. Flowered, "busy" dresses or blouses can direct viewers' attention to the dress rather than to the person.

- Direction. First-time talent needs more direction than does professional talent, especially about how to look natural. Have them look directly into the camera when delivering the information. Nonprofessionals have a tendency to look off-camera at someone in the room. They feel more comfortable talking to a real person than to a camera lens. This off-camera eye movement can look like insincerity. I often place someone next to the camera so that the talent has someone to talk to. Eye contact with the audience may not be best this way, but the more relaxed delivery may make up for that. If an interviewer is involved, be sure to have the talent respond to the interviewer, and where appropriate, make main points to the camera. Directing information to the camera keeps the audience involved. Also, be sure to inform the first-time talent that movements should be slow and deliberate. Sudden moves can be exaggerated by the camera. Quick moves can prevent the camera from following the action properly. Crossing legs, fixing eye glasses or adjusting in the chair can appear to be nervous gestures. Have the talent find a comfortable position in the chair before the tape rolls. If they must shift, fidget or adjust encourage them to do so between takes. A relaxed, natural appearance is what you are after. Informative communications with the talent will help create this illusion.

The first-time talent will often come to you before the taping session with questions regarding the production of the program, how they should prepare and what will happen on the day of the shoot. You should prepare a talent check-sheet to answer some of these questions for the talent you work with. *The Executive's Guide to TV and Radio Appearances* by Michael Bland (White Plains, NY: Knowledge Industry Publications, 1980), is a helpful book to recommend for a corporation whose executives make frequent television appearances.

The opportunity to work with professional talent is a rewarding experience. Professionals are usually easy to work with and do not bring with them the politics of the organization, reporting structure and inflated egos that often accompany executives who appear before the camera. Professionals usually become immersed in the production, not having other office responsibilities to worry about. They understand how important the program is to you and devote enough time to create the desired result. They do not guarantee that they will always get the lines right the first time, but they do guarantee consistency and compliance with the director's instructions. Professionals are aware of the details of production and how long it usually takes to get the scene right. They usually are understanding when it comes to the rigors of production.

The decision to use professional talent quite often is a budget decision. But before you make the decision, think through the situation thoroughly. Professional talent can save a great deal of time during production because they learn fast and can deliver lines believably without too much extra rehearsal time. They are able to look and move naturally, and often the entire production becomes more efficient. The time saved during production can often make up for the extra fees of professionals. The same goes for narration: professionals can read a script so it does not sound stiff, and they can add emphasis and excitement through trained voice inflections. Talent like this can add to the message, allow for correct and unbiased interpretation, and enhance the program. This enhancement alone often justifies the cost.

DESIGN RESPONSIBILITY

The medium of television appears to be a deceptively simple communication tool. In reality it is a very sophisticated medium that offers sound, motion and color to enhance the training experience. The director has to be aware of each element and its influence on the message.

The director should have the entire program envisioned in his or her head. Each sound that is important and that could alter the message must be thought through. In order to make its most significant contribution, each visual event, each color or motion that could affect the message must be understood in advance, thought through and executed.

The director really has to understand what will appear on the screen at any given moment in the program. The director has to control all of the dimensions of sound, light, color, motion and talent effectively, all during the program. During the production, the director should know which scenes and shots will be edited together so backgrounds, lighting exposure, audio, colors and action can be combined during the editing process and result in a flowing program that is easy to follow. The audience should not be jolted by a mismatch of action, exposure or audio change. An abrupt change of audio, for example, causes confusion and interrupts information flow. Likewise, the audience should not have to hunt for action on the screen or have to try to find important subjects. Backgrounds, exposures and colors should blend smoothly from shot to shot so that they do not draw attention to themselves and interrupt the message.

Transitions from scene to scene should be smooth, so the transition technique does not draw attention but allows the message to remain strong. All of the elements come together in a plan that, when executed in production, edits together smoothly.

CONCLUSION

Videotape is a visual medium; therefore, when you choose video to deliver training programs, remember that the content should lend itself to the visual experience. There should be good reason for using video to deliver any particular message and make it easier for the viewer to understand. Because video offers audio, motion, color and ease of distribution, it should increase the learning experience. But because it is a visual medium, and it does offer audio and motion, each of these elements has to be thought through in determining its contribution to the total message.

NOTES

1. *GTE Lighting Handbook*, 6th edition (Danvers, MA: 1978), p. 79.

2. Steve Floyd and Beth Floyd, eds., *The Handbook of Interactive Video* (White Plains, NY: Knowledge Industry Publications, Inc., 1982), pp. 2, 3.

3. Bernard Dick, *Anatomy of Film* (New York: St. Martin's Press, 1978), p. 33.

4. Herbert Zettl, *Sight, Sound, Motion* (Belmont, CA: Wadsworth Publishing Company, 1973).

5. *Audio-Visual Communications*, May 1984, p. 52.

8

Production and Post-Production

PRODUCTION—PHASE 4

The plans have been made, the script written, the talent rehearsed. Now it is time for the production to begin. Actually some of the production of the training program has begun already. If graphics are needed, that element has been put into motion; if slides are needed as inserts, they are ready; and if sound effects are needed during the shoot, they have been chosen and prepared. The production is the orchestration of all of the elements to create the sights and sounds needed for the training experience. It is time for the crew to capture the images needed and create, through the camera's eye, the message. (Fig. 8.1)

The goal of production is for the director to translate words on paper into images and sounds on the screen by means of the equipment, the crew and the talent. This effort amounts to only about 10% of the time devoted to the entire project. But it is an effort that produces all the elements that, when pieced together in editing, will become the end product.

A good director combines the roles and duties of an artist, a technician and a manager. The person approaching the job must stand ready to be a cheerleader, a nag, an arbitrator, a conductor and a juggler, and must not only perform all of these roles simultaneously, but perform them capably as well. The essential tools include the script, the talent, the crew and the equipment, in about equal order of importance. Tying them all together depends on the director's understanding of

97

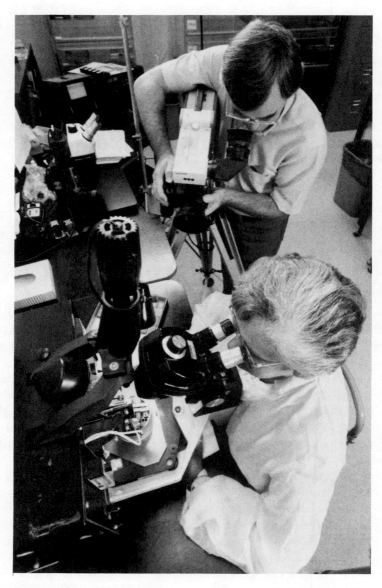

Fig. 8.1. Capturing intricate details is often necessary during the video-taping of a training program. This technique enables all the viewers in a large class to see exactly what is happening on the screen.

the process—the *gestalt* of production—and a dedication to
the program itself.[1]

The successful production is usually the result of a team effort. The
team, led by the director, operates video and audio equipment and deals
with props and talent to capture the right pictures required by the script.
Usually, everyone involved in production gets wrapped up in the
excitement and action. It becomes a contagious experience, filled with
action, fun and work, that results in a program that is seen by viewers
who in turn become involved.

This is where one of the problems of production lies. The viewers
are not present when the trade-offs are made. All productions have trade-
offs, the late afternoon decisions that seem insignificant at the time but
end up as gigantic mistakes in the final program. Obviously, the director
wants to make as few trade-offs as possible. Trade-offs can be avoided
by careful, detailed pre-production planning, but in the heat of any
production there are bound to be some. Guides to reducing the trade-
offs are:

- Nothing, absolutely nothing, replaces thorough planning. The
  details of the shoot, the equipment, cable, power sources, props
  and camera angles all have to be planned in advance. The energy
  in production should be devoted to getting the best possible
  performance from equipment, crew and talent.
- Go into the production with the program planned out and
  visualized completely in your mind. Know, through mental
  rehearsal, what is going to happen at any given moment. Know
  what the end product will look like.
- Take control. There is only one director on the set. There should
  be only one person making the decisions. Talent selection,
  location, mood, product and action should all have been worked
  out earlier with the content expert and the client. On the set
  there is no time for a client decision on what camera angle would
  show the product best. That is what the storyboard is for.
  Arguments should not arise as to what the talent should say.
  That is what the script is for. All major decisions should be
  made in advance, not under the pressures and time constraints
  of production.

- Display self-confidence. Your confidence in the equipment, crew and talent will be reflected in their performance. Challenge all of them to their limits but do not put limits on their capabilities.
- Work as a team. There should be only one director but each team member should be encouraged to contribute. The final decision comes from the director, but input by all should be solicited.
- Take detailed notes during production. This will help you during the post-production process. Notes should include such considerations as technical problems that might be corrected during post-production; what scenes, locations, shots and takes are on which tape and where; where last-minute shots are added; and where inserts and close-ups appear on the tape. Be sure to mark the reels well and note any narration or action changes in the script.
- Slate everything. A slate is a visual identification of the scene, shot and take. This identification, which appears at the start of each scene, will help you locate takes during editing.

During the production phase, all of the location shooting will be done. The shooting script and shot sheet will have been prepared so the director will follow the schedule to accomplish the shoot.

The location recording process is a complex series of events, and I would recommend reading one or several of the references I have listed at the end of this chapter to gain a greater understanding of the process.

All of the studio shots are also done during the production phase, such as close-up insert shots, graphics and slide inserts. The voice narration is also recorded at this time. It is always a wise practice to record the narration after the location video is done, because script changes may occur based on what was shot in the field.

POST-PRODUCTION—PHASE 5

The next phase is post-production. It takes place after the program has been videotaped during the production phase. The purpose of post-production is to assemble all the elements recorded on tape and edit them into the final product. But before the editing begins, some important steps should be followed:

- As soon as possible after the shooting is complete, reviewing and logging of the original footage should take place. Review the tapes to make sure that they are accurate and to ensure that everything that should have been shot was indeed shot. During the review process, take notes on what shots and takes look satisfactory so that an edit decision list can be created. This can speed up the editing process because many of the editing decisions will have been made during the review process. Detailed notes also prepare you better for the edit process by listing problems with audio levels, the location of insert shots, options for edits, and the places where close-ups appear for use later as inserts. Notes taken during production should be correlated with the review notes in preparation for the edit.
- As a result of what was shot in the field, an edit script may emerge. As you are shooting on location changes occur, and the script may vary somewhat from the original. A new editing script needs to be written based on exactly what was shot.
- Music and sound effects will be selected, edited and mixed on audiotape in preparation for transfer to videotape. It is wise to lay down on videotape prior to the editing session all audio and visual elements (including slides and graphs) to be used for editing. Music, sound effects and voice narration will be mixed together on audio equipment to create the final audio track. This will then be transferred to videotape, so the pictures can be edited to the audio track. Try to create the cleanest narration track possible because the program will be edited to this track.
- Think through and prepare special effects in advance. Storyboard and time the effect events so they will fit into the desired time frame of the audio edit. And be sure you understand what you are saying with the effect chosen. As all effects add messages of their own to the program, you must have a clear understanding of what you are adding with these effects.
- Always think through the editing event. Know where you are going to next in the script and how you are going to get there.
- Take careful notes during the editing process. Audio levels and technical settings change over several days of editing.

As with film editing, the original footage shot on videotape is usually out of sequence. The editing process will take the original footage and place it in the correct sequence according to the requirements of the script. This process is both a mechanical function of the equipment used and a creative process of the editor. The editing machines electronically copy the original footage from a playback recorder onto an editing recorder, following the commands of a series of buttons. The act of pushing the right buttons is relatively simple and can be learned in a few hours. But the creative process, deciding what shots to use, in what sequence and at what pace, is more difficult to learn. "The amount of control a video editor exercises over these decisions differs from production to production. In dramatic productions, video editors often work from a script marked with the producer's or director's instructions. When this is the case, the editor usually has very little control over the selection and sequencing of shots."[2] In meeting most of our organizational training requirements, the director of the program will usually do the editing as well. So the creative process of shot, sequence and pacing will be initiated by the director. The decisions made will influence the final product. The editor sets the pace for information delivery, selects the shots that are most appropriate for the script and sequences the program to meet the learning requirements. It then becomes obvious again how important it is to know what the program must look like so the editing of the video and audio will achieve that end.

The editing process usually takes two or three attempts to reach the program objectives accurately. You will probably find yourself reediting a program several times to meet the goals of the client or content expert. You can add to and enhance the training through editing. Not everyone is going to agree with your editing choices. Because there are many editing choices in a typical program, it is best to plan on several editing sessions to reach the desired and accepted result.

After final edit is made, and depending on the editing system that you are working on, the character generation, music and effects may be added at this time. Often the rough edit made for the approval process does not contain effects or music. These are added after the rough edit has been approved. The rough edit will create the pacing and timing of music and effects. Once approved, these effects will be placed in the proper portion of the program.

DISTRIBUTION

The distribution step contains details that are often neglected until the last moment but do require planning. Copies of the videotape program will have to be made for the audience. Dubs are made in real time, so a twenty-minute program will take twenty minutes to copy. This time factor can become significant if twenty or thirty dubs have to be made. Along with the dubs the program labels will have to be prepared. Any support materials such as student workbooks and instructor's guides will have to be prepared and labeled. The entire training package will then be prepared for mailing, if the program is to go out to field offices. The package that holds the tape, printed materials and perhaps slides or transparencies will have to be designed and manufactured. You should give thought to this package and how it will actually be received by the audience. As mentioned earlier, the impact of the entire training "package" will influence its reception. If the training materials and the package are well-designed and easy to use, the training experience will be better received. Several examples of well-designed total training packages are shown.

These packages are designed so that the viewer can easily identify the training title and the program number, if it is a series. The workbooks, tapes, and student and instructor guides are clearly identified. All of the graphics are consistent, carrying a strong program identification "look," and reflecting a continuity of design.

Arrangements for participation in the training event are usually made by the training department, but you must give thought to arrangements for the viewing of the videotape. Playback equipment may have to be scheduled or rented.

NOTES

1. Lon McQuillin, *The Video Production Guide* (Santa Fe, NM: Video-Info Publications, 1982), pp. 211-212.

2. Gary H. Anderson, *Video Editing* (White Plains, NY: Knowledge Industry Publications, 1984), p. 9.

9

Resources: Equipment and Staff

As you have learned in previous chapters, the bulk of the work devoted to developing video training programs is the pre-production phase—the design, scripting and logistics. Seventy percent of a producer's time is devoted to the pre-production phase and only 10% to actual production. The remaining 20%, on average, is spent in the editing or post-production process.

Based on these averages, the trend in the industry is to get away from investing in large video production facilities and elaborate equipment, and invest in the creative talent of producer/writers instead. Similarly, aware of the time devoted to the post-production process, organizations are devoting more resources to the purchase of editing equipment and renting whatever production equipment they need for a given project.

This chapter is intended as a guide to selecting equipment and staff. It is based on current industry trends. I will first emphasize the importance of conducting a needs analysis before investing in equipment, and then I will offer my recommendations for equipment. I will then look at staff selection considerations.

THE IMPORTANCE OF A NEEDS ANALYSIS

As I mentioned earlier in the book, video is most often used in organizational television for the purpose of training. Therefore, if television fits within the training objective of your organization, purchasing equipment may be your next step. Before you hire any staff or purchase any equipment, however, you should conduct a thorough needs analysis within the organization. The specifications for the equipment should be defined by the needs of the organization. You have to look at the big picture; the training needs of the organization, the resources that can be devoted and the trends of the industry. Begin by asking: who are the clients that will be using video? Where are they located? What are their individual needs? What types of programming will they require?

A thorough needs analysis will include interviews with potential users of video to determine the types of programming needed, the locations of the potential audiences and the ways you are to reach these audiences with your programs. And you have to know what resources (staff, time, facility and budget) will be made available for video. These questions and many more will help create a plan for the development of video in an organization.

Three influencing factors should be addressed in the needs analysis: convenience, confidentiality and cost.

To what extent will the convenience of having video production staff and resources available at the organization influence the decision? Will corporate staff have to travel off-site to a production company to deliver video messages? Will it be inconvenient to take products off-site to have them recorded? Having video on-site, available to executives, may be an important consideration for your organization.

Confidentiality may be important, particularly when it comes to making product information tapes. It may not be prudent to let information leave the organization to be recorded at an off-site facility. This is an important issue for law enforcement, health care and high-tech industries. If organizations deal with sensitive material, it may be required that in-house resources be available to produce the desired information.

When considering cost, you may find it cheaper in the long run to produce videotapes in-house than it is to go outside. As I have noted earlier, the average in-house videotape program costs about $6000 to produce. The average outside production costs anywhere from $1000 to $2000 *per finished minute*.

The needs analysis should show that the commitment of resources for video will benefit the organization. This benefit should reveal financial gain. The results of the programs that are produced on the equipment you purchase should far outweigh the resources devoted to setting up the facility. The growth of the facility should then depend on achieving success with the programs that are produced. In other words, the facility should provide a high return on investment.

If you are just beginning to set up a facility, you should consider acquiring an independent video consultant to help you with your plans. Investing in an experienced consultant up front may save you thousands of dollars in the long run. An independent consultant has no equipment to sell and will analyze your needs objectively.

VENDOR SELECTION

One of the first concerns in setting up and establishing a video facility is developing a working relationship with vendors of the equipment. Keep in mind that you should always be in the position of telling the vendor what you need, not the vendor telling you what you need. As a staff member, you know best what video facilities your organization requires. The vendor does not know the politics and the budgets, the staff, the space, the facility or the talent within your organization. The vendor does not know your program requirements and limitations, production demands, priorities and restrictions. You do. You have the answers based on your needs analysis. Before going to a vendor to purchase equipment, you must know your needs. Your should dictate to the vendor the equipment you need in order to put your package together.

Points to consider in vendor selection include the reliability of the vendor. The vendors should supply you with names of customers they have served over the years. Call up a few and get their feelings about the vendor in question. Find out what kind of support the vendor supplies. Look at the repair or service area. The vendor should provide in-house service and be certified to handle the equipment you are purchasing. Walk through the service area. Would you want to take your equipment there for service? Check out the parts supply. A reputable large vendor should have a suitable stock of parts for all the equipment you are purchasing. Being able to fix a piece of equipment is one thing; it is quite another to have the right part to fix it with.

Talk to the vendor at length about the guarantee and the warranty of the equipment you are purchasing. Find out if the system will be hooked up and tested at delivery, and if the vendor will supply you with equipment training.

Furthermore, if the equipment does break down, will the vendor service it at your facility or will you have to bring it in? If so, will the vendor supply loaners while your equipment is being serviced?

It is essential to develop a solid relationship with the vendor. You are going to invest a great deal of money. Be sure you know as much as you can about whom you are investing it with.

A word about drop box purchasing. A drop box purchase is getting the best price out of a warehouse in a large city, having the equipment delivered by air freight and dropped at your receiving dock. You may get the best price, but you have no guarantee that you are going to get any kind of service or support for the equipment when you need it. I do not suggest shopping just for the best price. Video equipment is very sophisticated electronic gear, and it does need technical attention. Over the years it will need reliable service support. So it is usually best to buy it locally, if you can, to receive the support you will need. Saving $50 by buying a drop box out of New York may, in the long run, cost you hundreds of dollars in service.

EQUIPMENT SELECTIONS

It is not my intention to compare cameras, citing specific models, manufacturers and costs. New equipment appears on the market so fast that specific model numbers would probably be outdated by the time this book is published. I will attempt to define cameras, recorders and equipment in a broad sense, and you can place specific model numbers within that framework based on your suppliers, budget and individual needs.

I would like to present the equipment discussion by suggesting three stages of equipment purchase. I will first outline a minimum or initial equipment package. I will then suggest an upgraded package, and an optional equipment addition. Always keep in mind that any actual equipment purchase should be based on your thorough needs analysis. These equipment packages represent typical video systems, and I offer them only as a guide.

What I am suggesting as the initial expenditure for equipment is a conservative, minimal package. After you use the equipment for a year and produce quality programs that meet your objectives and show a satisfactory initial return on the investment, you can move into a second phase or upgraded package.

The equipment package will be broken down into three groups: production, editing and distribution.

The current standard production format for producing video training programs is 3/4-inch U-Matic videocassette. If you are seriously aiming for video results within your organization, the format to choose is 3/4-inch U-Matic. A great deal is being written about consumer 1/2-inch VHS, but as an experienced producer who has worked in both formats, I would recommend going with the industry standard. It is a superior format and will provide superb results for the money invested.

A highly regarded engineer in the video industry, Cecil Smith of Joiner, Pelton and Rose, an engineering consulting firm in Dallas, agrees:

> Although attractive in price, the video signal quality of multiple-generation tapes produced on 1/2-inch VHS or Betamax format still does not measure up to the quality afforded by 3/4-inch equipment. The picture and sound reproduced from original production tapes using either format are good. Second generation 'edit master' tapes using 1/2-inch equipment are fair, while 3/4-inch tapes remains good. Third generation tapes, normally used for 'release prints' that are seen by the audience, are poor using 1/2-inch tapes but remain fair to good with 3/4-inch tapes.
>
> The quality of the signals and the life of the equipment are directly proportional to the cost of the equipment. The higher the cost, the higher the quality of the signal or the longer the life of the equipment. In short, you get what you pay for.

Production Equipment

With this cost qualification in mind, I would like to suggest a low-cost television production package that could be used for recording events, lectures and demonstrations, and for creating simple training programs:

- *Single-Tube Television Camera System,* including lens, batteries, battery charger, AC power adaptor, cables, fluid-head tripod, test charts and carrying cases. A typical low-cost camera system designed for training applications generally runs from $4000 to $6000.

• *Microphone and Audio System*, including microphone holders, cables and carrying cases. Typically lavaliere clip-on microphones are used in low-cost production packages, and the signals are fed directly to the microphone inputs of the videotape recorder. A typical low-cost microphone and audio system for training applications generally costs from $200 to $300.

• *Videotape Recorder System,* including batteries, battery charger, AC power adaptor, cables and carrying cases. A typical 3/4-inch U-Matic portable videotape recorder system for training applications costs from $4000 to $5000.

• *Lighting Equipment* is essential to the generation of pictures of reasonable quality from any television camera. A typical low-cost lighting equipment package for training applications generally has three fixtures and costs approximately $1500.

An upgraded version of this system that would stay in the 3/4-inch U-Matic format but yet bring the signal up to broadcast quality would be:

• *Three-Tube Television Camera System,* including all the accessories and support equipment listed for the single-tube television camera. In general, a three-tube camera will give color television pictures with more color saturation and more color detail. A typical moderate-cost camera system designed for training applications generally costs from $10,000 to $15,000. Higher-quality camera systems can go for as much as $40,000 to $50,000.

• *Microphone and Audio System* for signal quality and production flexibility can include several microphones (with accessories) and an audio mixer to combine the signals. The lavaliere microphones described above in the single-tube system could be enhanced with narrow pickup pattern shotgun microphones or wireless microphones. A typical moderate-cost microphone and audio system for training applications generally costs approximately $2000 to $3000. There can be a wide range in this budget, depending on the complexity of the desired equipment complement.

- *Videotape Recorder System,* including the accessories listed above in the low-cost system, should produce broadcast quality on 3/4-inch U-Matic format and will allow for considerably more production capability and flexibility than the industrial version with time-code capabilities. A typical videotape recorder system for training applications generally costs from $4000 to $6000, although this figure can go much higher.

- *Lighting Equipment* is just as essential for the generation of quality pictures from a three-tube camera as from a single-tube camera. The low-cost lighting equipment package detailed above can be enhanced on a piecemeal basis. For example, a large softlight can be added to the package to increase the overall lighting level in a scene. Generally, the basic lighting equipment package detailed above for the low-cost system can be enhanced for about $1000 per fixture.

Editing Equipment

A basic editing system which would complement the production package and allow for cuts-only editing without special effects would be:

- a "source" video player
- an editing video recorder
- an edit controller or "editor" to control the functions of the player and recorder automatically
- color monitors for the player and recorder
- audio monitors (amplifier and speakers) for the input and output of the player and recorder
- a character generator for placing titles on the picture

This level system typically costs between $14,000 and $18,000.

An upgraded version of this editing system that would allow for dissolves between two source machines, time base correction to stabilize the signal and enable the dissolves to take place, audio mixing and processing, and waveform monitor and vector scope to monitor the signal would add approximately $30,000 to the cost.

A natural next step in system advancement for both editing and production is the acquisition of time code. Time code is an electronic, digital signal that is recorded on the audio track or cue track of the videotape allowing for accurate identification of each frame of video recorded.

Most basic editing systems reference the control track that is automatically recorded onto the tape. Control track referencing counts pulse as the tape is shuttled from an operator-set zero. This count is displayed in minutes, seconds and frames. The control track allows for a relative tape position with a reference that must be reestablished if the tape is removed from the machine or if the machine is turned off. This is not a very accurate way to reference edit points.

Time code, however, gives an absolute tape position that travels with the tape allowing for very accurate editing. The difference between control track and time code editing is more than the concept of referencing. In general, time code equipment is considerably more expensive, but it gives a greatly expanded flexibility in automated editing procedures.

To place time code into the recording and editing process requires the acquisition of a time code generator, reader and display device. The edit controller will have to have the capability of reading and responding to time code. Adding time code capability increases the cost of the system significantly but allows for greater speed and accuracy in the editing process.

Further Considerations

- It is generally accepted that you should budget from 7% to 10% of your total capital equipment expenditure for maintenance.
- Budget $200 to $300 the first year for tape.
- Consider consoles, furniture and racks for your editing system.
- Budget for storage cabinets for your tape.
- Budget for cables, connectors, lamps, and extra batteries for the recorder and camera.

- Budget for travel cases for your production equipment if you plan to do any traveling with your gear.
- Budget for a cart to move the production equipment around.
- Optional items are music library; small, portable monitor for the field; and portable waveform monitor.

You get what you pay for when it comes to purchasing video equipment. And to reemphasize what my experience and the experience of the producers interviewed for this book suggest, for signal quality and production flexibility go with the current industrial standard: 3/4-inch U-Matic videocassette. But to duplicate the production equipment listed above in the consumer 1/2-inch format you will need to budget only about $5000. To select a 1/2-inch editing system you would need to budget approximately $7000. (Fig. 9.1)

A WORD OR TWO ABOUT SPACE

Once you buy the equipment you will need to put it somewhere. Unfortunately, space considerations are often overlooked. Some very specific requirements are needed for the housing of videotape equipment. A direct correlation exists between adequate video facilities and the ability to provide satisfactory service. Space commitment must be completed before you purchase equipment because architectural modifications of allocated space will have to be made. Each organization will have different space limitations, so it is difficult to generalize about video space requirements. At best I can offer some suggestions for housing the equipment, but part of the initial needs analysis should detail facility requirements for each individual organization.

Again, I would recommend consulting with a qualified, *video* engineer before planning the facility. I stress video engineer because most of the architects and designers of office space do not understand the unique requirements of video production. I can cite numerous examples of being called in personally to consult on a facility *after* the architects had designed and built the space. The carpet on the studio floor, dark wood paneling on the walls and electrical outlets at the customary six inches above the baseboards looked fine but proved to be wholly inappropriate for video production.

Fig. 9.1. A typical 3/4-inch U-Matic videocassette editing system with player, controller, recorder, waveform monitor and character generator.

Acoustical, electrical and air-conditioning requirements are only a few needs that must be planned for properly. Here is a sample of other requirements that you should consider:

- Production (studio) space and post-production (editing) space should be separate and generally should be located away from high noise and traffic areas, and on the ground floor for ease of equipment and prop access. Separate office areas for staff and separate areas for storage and maintenance should be assigned.
- Production space: 25′ × 30′ would be a minumum size requirement.
- Ceiling height: 12′ is minimum for lighting grid and ventilation.
- Extra large vents of air circulation, along with fiberglass (not metal) duct work for sound control.
- Extra air-conditioning for both studio and post-production area, individually controlled.
- Vents in both studio and post-production area placed properly for equipment cooling and sound control.
- Sufficient power for studio and post-production area (at least 100 amps for the studio and 60 amps for the post-production area coming from outside lines with no other power requirements placed on the lines).
- Avoid door thresholds for ease of equipment movement and avoid carpet and tile on studio floor. Carpeting is good in the post-production area for sound control.
- Floating floor for post-production area for ease of cabling and maintenance.
- Ample space should be devoted to the post-production area because it will house editing equipment, special effects, audio sweetening, camera inserts and duplication. Pay special attention to the environment being created. Use light colors, carpeting, drop lights to avoid reflections in monitors, comfortable seating for operators and clients, and good acoustics for sound playback and control.

There are many more detailed considerations for effective space planning of the production area. This list should emphasize the importance of obtaining expert help in facilities planning.

STAFFING REQUIREMENTS

The single most influential factor for any training and media facility is staff. Creative staff can make even the smallest of facilities effective. The first consideration in starting up any production facility is the hiring of qualified staff. Equipment won't create the programs; people will. Hire qualified people to produce the needed programming. Your first concern should be to develop programs that meet training objectives. This development is accomplished by producers and writers who have the ability to mold content into effective video programming. The operation of the equipment is only a small part of the entire program development process. Too much emphasis has been placed on hiring technicians to operate the equipment and not enough on hiring talented program developers. Organizations often fall into the trap of buying equipment first and then hiring a technician to operate it, without thinking about where the program ideas, design, concept and scripts will come from. Quality programming doesn't materialize magically when the camera is turned on. Talented people make it happen.

The following are general staff requirements for a new facility. Obviously, based on the original video needs analysis, staffing requirements will vary from organization to organization and different emphases will be placed on the positions based on the resources the organization already has in place.

Producer/Writer

For a new production support facility for training, the first person hired should be a producer/writer who has the talent to design, write and develop the video training program from start to finish. This person should be an experienced producer who will acquire the necessary resources to put the program together.

At the appropriate time in the evolution of the facility this producer will become a manager and handle all the management needs of the facility. He or she will then turn over the producing and writing responsibilities to other staff as the facility grows.

Your first priority is to create effective programs. If you develop an appetite within the organization for video, you'll find that equipment and technical support will follow.

The experienced producer will command a salary of $25,000 to $35,000, depending on the part of the country the organization is located in. This person should have the talents to:

- Assess the video programming needs of the organization
- Design, write and produce quality programming
- Assess the budget, equipment and resources needed to carry out production requirements
- Supervise production staff as required for productions and as facility grows
- Evaluate program effectiveness

Video Manager

Very soon in the evolution of the facility, management skills will be required. Organizations then hire video managers with specific management experience, or the producer evolves into the facility manager.

Video management responsibilities include:

- Planning, organizing, budgeting, staffing and administrating the facility
- Writing policy and long-range plans for the facility
- Recommending equipment purchase, staff and facility expansion to upper management
- Developing and recommending technology advancements to meet overall organizational communications objectives
- Supervising staff development

Experienced video managers command salaries of $30,000 to $50,000.

Production Assistant

The next staff member will be a production assistant with specific technical skills for equipment operation. Some areas of responsibility would be:

- Maintaining and operating production and editing equipment
- Assisting in arranging for production logistics
- Assisting in logging and reviewing production tapes in preparation for editing

- Duplicating program tapes in support of distribution network
- Setting up, operating and maintaining playback equipment in support of organization training efforts
- Maintaining logs and schedules for equipment maintenance

As the production requirements grow within an organization, the manager should consider hiring free-lance talent before hiring full-time staff. Hiring full-time staff burdens the organization beyond just salary. It is generally accepted in industry that a staff member will require double the salary in support services and benefits. So it is always wise to analyze staff requirements carefully before committing to full-time positions.

Other positions that will need to be filled as the facility and the services grow will be technicians, photographers and graphic artists.

In setting up a video facility in support of the training function you should first do a thorough needs analysis. Create a plan that will frame equipment purchasing, staff support and facility construction. Start small; create an appetite within the organization for video programming. Evaluate the results and grow as the return on investment is realized through successful videos. This phased-in plan is a sensible business approach.

Chapter 10 will become a valuable guide for analyzing staff and facility resources in support of various video training responsibilities. The managers interviewed for that chapter are some of the best in the country and have developed their facilities to meet the needs of their respective organizations.

The International Television Association publishes a yearly salary survey that breaks down salaries according to regions of the country. It also contains brief job descriptions for the television industry and typical operating budgets of its membership.

DISTRIBUTION

After the video training program has been designed, shot and edited it has to be seen by the target audience. This can become a big problem for organizations. The best produced video training program may not work because there may not be a system set up for the audience to see it. A training program's success depends to a large degree on a successful distribution system.

When I consult with various organizations on setting up video for training, one of the first questions I always ask is, "How will the audience see the program?" Ask this key question before you even plan the program. The answer to the question may indeed affect how you produce it and in what format.

In workshops I do around the country on setting up a video training facility, I always ask the audience, "How many of you have an effective distribution system set up within your organization?" Of the 50 or 60 people in each of my audiences, only one or two raise their hands. Why is it that so much time and effort is put into designing, scripting and producing a program but so little into figuring out how the program will actually be seen by the audience? Programs do not just end up in a playback machine in a field office, turn themselves on automatically and play back to the audience. You have to think through the management of your distribution system.

When planning for the development of a video training program, you should include distribution costs in the budget along with design and production costs. Distribution costs include tape, dubbing charges, packages for mailing, printing for labels, mailing, scheduling, bookkeeping, cleaning, repair and cataloging. This distribution system has to be maintained, and time, effort and personnel have to be devoted to it.

Other distribution questions to raise are: Will customized mailing packages be designed? What about the materials that will accompany the videotape? Who is responsible for the printing of the materials, binding, packaging, mailing, distribution and labeling? Am I going to create a catalog of programs, and how am I going to promote my programs and my catalogs? What about an evaluation or feedback system? How am I going to know who saw the program, or how many people saw the program? How many showings did the program have in any one location? How am I going to find out the general reaction from the audience? Will the field offices create their own videotape libraries?

How many locations need playback equipment and tape? What's the geographic spread of locations? Geographic spread will help decide on timeliness of distribution and dissemination of information. You will have to consider the ease of obtaining the hardware for playback and the reliability of hardware in each of those locations; the number of companies providing support for your hardware in the field locally, which leads to accessiblity; and the cost of maintenance locally. Who will be using the hardware in the locations and who will be responsible for the hardware? Am I going to have a field coordinator who will keep the maintenance logs and keep the equipment running? Am I going to provide backup loan machines when field equipment breaks down?

What commitment is to be made in support of the playback network? What resources will be required in terms of my responsiblity and management of it? Will I have to train people on the use of the equipment in the field?

The economics involved in setting up a large network for distribution will have to be evaluated and the commitment by the organization will have to be made so intelligent and effective network management can take place.

There are no pat answers about distribution systems. Each organization must meet its own delivery requirements. The four distribution systems most commonly used to deliver training programs present a variety of options and advantages.

Videocassette

VHS has become the leading distribution format for organizations because of the low cost of its playback equipment and the low cost of tape. Another factor contributing to this format choice is the consumer demand for VHS. VHS is selling some 80% compared with 20% for the Beta 1/2-inch format. Corporate video managers are setting up networks in response to the high consumer placement of 1/2-inch VHS. Having so many VHS machines out there in the field almost guarantees easy access to corporate programs.

Managers I have interviewed for this book noted that their 3/4-inch U-Matic machines are being replaced in the field with 1/2-inch VHS. More and more companies are turning to this format for their videocassette networks.

Interactive Video

Interactive video is an exciting medium that is going to touch all of us who are involved in training. Considerable advancements have been made in interactive video in the last couple of years. Cost reduction in the production and mastering of interactive programs has made it a realistic choice for many training situations. We will see interactive video playing an increasing role in training in the years to come.

Interactive video becomes an extremely effective training tool and distribution system at the same time. It has the ability to get the student involved in the learning experience and offers training management and program evaluation. Interactive video programming allows the viewer to control the amount, sequence and pace of the information being presented. The interactive program design becomes the teacher of the material being presented. It reacts to, encourages and rewards the participant. It allows the learner to make judgments and take actions. It lets the learner explore options to see what can happen in a given situation. Wilson Learning Corporation, along with many other training companies, is devoting more resources to interactive video for these reasons. As Linda Antone, Vice-President of Video Production, explains, "We are in the computer age. The video and computer games that are so popular today cross over to the training experience. The new generation is more actively involved in education . . . more experience-oriented. Videodisc technology allows for active learning to take place."

This active learning happens because of the almost instant retrieval capability of videodisc. Digital programming information is stored on the disc with 54,000 frames readily available at the push of a button. "Pages" of words, charts, graphs, animation or full-motion video can be retrieved randomly within seconds. Most interactive programs are designed so that the viewer must react to the material and get involved with the teaching process. This creates truly individualized instruction.

Furthermore, because of the nature and flexibility of the equipment, interactive video allows for training management to take place. It enables the training manager to monitor student response, and provides testing and certification of materials. It can evaluate the performance of both the trainee and the training system.

Large videodisc networks are going into service every day; Apple Computer, Atari, Bally Corporation, Coca-Cola USA, Cuisinart, DEC, Ford, General Motors, IBM, Merle Norman, Miles Laboratories, Minolta, Sperry Corporation, Toyota and Xerox Learning Systems are just a few of the large companies that are turning to interactive video to solve training problems. This commitment to the technology will bring equipment and programming costs down. It will also make more generic programs available to allow smaller organizations to justify the investment in the technology for training.

Toyota Corporation is an interesting example of a large corporation committed to video training. According to Jim Worrel, Video Program Administrator, Toyota chose both interactive video and videotape for its distribution network. Its video network consists of approximately 1200 Pioneer, Level Two, Laserdisc players and JVC videotape players.

The network reaches over 20,000 people. The programs are produced for corporate communications, training, sales and service.

Toyota clearly had two types of programs that needed to be distributed—fast turn-around, corporate communications with a relatively short shelf life and more sophisticated training programs that could be used for several years. For the communication programs, videotape was chosen, and videodisc became the medium for training programs in which interactive programming and long shelf life are required.

Toyota represents a unique example of building a video network that reflects the needs of the organization and addresses programming considerations. Each video format has different characteristics. Toyota chose to combine formats to meet the needs of the network.

With 8-inch videodisc technology being widely introduced now, the economics of using videodisc in our training applications becomes easier to justify.

Computer-enhanced Video

Linked with interactive video systems, the personal computer is becoming a dynamic teaching tool, poised to make a significant, impact on the training environment. If I were a gambler, I would place my money on computer-based training (CBT) in the next couple of years.

Computer-based training systems are highly effective and economical tools to develop and distribute individualized training. According to an authority on CBT, G. Phillip Cartwright, PhD., Director of Research and Training Laboratory for Technology, Pennsylvania State University, and creator of the first full-length computer-based training course in the nation:

> CBT reduces training time and adds significantly to individualized learning. It offers a great flexibility in tailoring content to individual student needs, and can be modified easily for student loads and populations. Because of the computer's large capacity to store information, it teaches, schedules, tests and evaluates training activities. CBT becomes a training management tool for accumulation of student profiles, class lists, registration, test scores, evaluations and schedules. And it analyzes performance data and adjusts curriculum based

on student responses. It is clearly the training delivery medium of the future.

CBT is an active learning process that allows students to acquire new skills, information and concepts. It is very effective for drill and practice, and often delivers the training on the very equipment the student will be operating on the job.

Authoring systems such as Phoenix from Goal Systems in Columbus, Ohio, and ScholarTeach from Boeing Corporation in Seattle, Washington allow trainers to design training programs on working main-frame systems in use throughout the organization. In this case the training delivery system is already paid for and in place. A tremendous potential exists for reaching virtually everyone in the organization who works on or has access to a terminal. The cost of training and training delivery is justified through the number of participants the programs can reach.

Typically, information is presented to the student by the computer, and the student is asked to respond to a question or react to a scenario posed by the computer. The student then makes a response through an input device such as a keyboard, light pen or touch screen device. The response is analyzed by the computer, and the student is given more information and asked to try again, repeat information or present alternate versions, etc., all dependent upon the student response. The computer maintains records of the responses and guides the student through a virtually infinite number of possible paths in the mastering of learning objectives, depending on the student's strengths, weaknesses, partial knowledge and current performance.

The research shows that CBT works well in meeting training objectives. The computer, when coupled with videodisc, creates a powerful training tool that delivers cost-effective training. This is a growing technology that all trainers should follow and place in their arsenal of training tools.

Videoconferencing

Immediacy of information, numbers of people reached and reduction of travel costs are the three leading factors that make videoconferencing an attractive training delivery system.

Large numbers of people can have immediate access to timely information through videoconferencing. It opens greater numbers of people to information. Instead of sending only one or two people to a conference, you can transmit that conference through videoconferencing to hundreds in need of the information. And since videoconferencing is based on transmitting a live performance, that information is current. With the further capability of two-way audio and video, the training event enables the experts to present the information and exchange ideas with viewers locally. Conference participants throughout the country have interchange capability opening up tremendous channels for local input.

Jerry Freund of Tucson Medical Center says,

> We have seen in the last three years the cost of the satellite equipment dropping by 40%. This cost reduction will allow smaller organizations to purchase satellite dishes. This will then allow the organizations to participate in national conferences and will make experts in their fields of interest available to them with two-way interchange.
>
> Costs of transmitting the information are coming down as well. With this we will see more dedicated networks created, smaller groups can take advantage of the networks and the information being presented can become more localized. This will also lead to more local input for planning and designing the conferences. This is the area I feel will really make videoconferencing attractive for the trainer.

Throughout the country last year more than 500 videoconferences, representing more than 2000 hours of live television, were held. Sixty percent of the networks set up are for corporations, with 40% used for training. Organizations like Hewlett-Packard with approximately 65 downlink sites and a hundred more planned are using it for training, marketing, sales events and corporate communications. The Voluntary Hospitals of America dedicated videoconferencing network reaches 90 hospitals at a time, delivering current hospital legislative issues, management updates and development, board meetings, medical updates and educational programs.

> Hailed as the medium of the eighties, videoconferencing is a communications tool that combines sight and sound to bring together groups of people gathered at any number of locations into a single, 'electronic' meeting. Satellite videoconferencing

differs from other communications media in that it uses full-motion, full-color, live video images. It is, in fact, live television.[1]

NOTE

1. *The Producers' Guide to Satellite Videoconferencing* (Atlanta: Video Star Connections, Inc., 1985), p. 1.

10

Video Training Case Studies

In this chapter we look at five very different applications of video training and hear from some of the best video training producers in the country.

UNION PACIFIC RAILROAD

Alexander Tice is Media Production Supervisor for the Union Pacific Railroad. He has produced interactive video programs on company history and an introduction for the Union Pacific museum and visitors center in Omaha.

Applications

The company also has many day-to-day applications for video. "We use it in a number of ways fairly typical of the way everybody else does," explains Tice. "We use it for management communications, safety, training, problem solving and for general information throughout the company."

A new product or service can be introduced to customers or to employees easily through video. Union Pacific had just such a new service, and the company realized that the only way it would work was through employee understanding. Tice had the answer with video:

> In our customer service center, which has recently been computerized, we have staff who have never before dealt with a customer who are now assigned to handling calls from the

127

outside. They're not familiar with some of the questions people are calling in about . . . how they can trace a car or a shipment. This information is in the computer, but some people just aren't up to speed with it.

The video program had to make the computer understandable to employees as well as explain to them how to deal with customers. Otherwise, Tice says, "The customer would go somewhere else. Basically it was a program on telephone courtesy, how to handle customer calls, why it's important to deal with customers properly and how to get information. The program is being used in all the company service centers, with backup training from managers."

Union Pacific has a well-established distribution system. "We master one 3/4-inch and make duplicates on 1/2-inch VHS," says Tice. Field offices then use less expensive 1/2-inch format machines for playback.

Another area where video plays an important role for the company is in problem solving. Tice recalls a very costly problem that his department was able to solve with a training tape:

> During a spell of extremely cold weather, the railroad suffered an awful lot of damage to its locomotive fleet because of fuel line problems. The problems caused the engines to shut down, and when the engine shuts off in sub-zero weather, the water lines freeze, and you suffer physical damage to various parts. This represents millions of dollars in damage in a very short period of time.

Audio Visual Services produced a videotape on the problem, documenting the damage and the losses. The production team visited the company laboratory to show how the problem actually happened. A new fuel mixing device was developed that solved the engine shutdown problem. But the video department was still not finished.

> We took all the information, showed the procedures on how to shut down a locomotive in cold weather and did a role-play between the engineer and a dispatcher on what to do about the problem. Soon after the program was produced, we had another cold spell that could have resulted in a large

amount of damage—a half a million dollars worth—but as a result of the program the losses were only $50,000.

Equipment and Staff

The Union Pacific video facility consists of a studio measuring 20 × 30 feet, minimally equipped, because Tice says most of his work is done in the field. For electronic field production (EFP) he has Hitachi SK-80 and SK-91 cameras as well as Sony BVU 110 and 4800 field recorders. For editing purposes, a full Sony 800 editing system is used. The company also has a film chain for slide and film-to-videotape transfers.

Tice says improved quality has taken time:

> We didn't start at the top and get the best. We started small and grew over a period of time. I'd like to stay with 3/4-inch for a period of time because of all the work that has been done that way. But one of the reasons we went with BVUs was with an eye towards going to the Sony BetaCam, which can be integrated with the 800 editing system.

Tice supervises four producer/directors who turn out an average of 65 completed programs a year. The cost per program can range from $1,000 to $5,500, and sometimes higher because of special effects that must be done outside the facility.

Tice is optimistic about the future:

> We've had a lot of growth in video training. On-the-job training is very important as people are shifted from one job to another within the company. As the corporate needs change, instead of hiring different people to meet those needs, we would rather retrain someone already on the payroll. Video helps train our people at a reduced cost. And that's important.
>
> As far as what will happen in the future here, I really believe that interactive video is going to become more important. You can demand more from the student, getting him or her to participate in the process. And you can monitor how fast or how well a student is learning.

The "Sell-in" sales training tapes had reached the grocery stores and made a Coors sales representative's job easy. "Buyers just give them (sales representatives) five minutes, then it's out the door," he says. "The five-minute tape was very effective."

Coors is also enthusiastic about communicating with and training its employees. Kultala says his department produces management training, skills training and safety training videotapes on a regular basis as needs in the factory change:

> People are important to the company, so we don't necessarily lay workers off, just reassign them and retrain them. And safety training is heavily stressed. We produce quite a few safety tapes for our employees. They can range from "Safe Operations of a Fork Lift" to "Proper Brewery Systems Cleaning." Employee communication programs are used to explain health benefits or health awareness or how to deal with stress. It's good internal PR . . . As far as a break-down on whom we produce for, right now I would say it's a 50-50 split between training and corporate communications.

Kultala is also very pleased with an employee educational series his department produced:

> It's a series on economic awareness, free enterprise and economic issues initially intended for our employees. That series has now been sold to other corporations, and we've more than recovered our production costs. It's also put Coors in a position of being an authority on economic issues.

Some of the more routine tasks for the production department include time shifting of classes (recording presentations for playback at a later, more convenient time), documentation of problems or big events in company history and providing television news departments with video of the brewery or the Coors family.

> We keep finding out that slides and overheads do not have much impact and you do not get the message across unless you have one heck of a presenter. Even the really good slide/

tape shows we have are transferred to videotape for convenience. When we look at a new production we automatically look toward video as the first choice.

Equipment and Staff

Kultala oversees a staff of eight in a brand new studio. He says high-quality equipment is important to the operation. Field work is done on a BetaCam 1/2-inch tape with post-production on 1-inch. Most of the distribution is done on 3/4-inch tape.

The cost of a finished minute, according to Kultala, varies, but it is averaging less that $500. "Our budget is half a million dollars a year, and after we totaled up our costs last year, we figured it would cost twice as much to do what we did using an outside production company."

Kultala is looking forward to the future at Coors: "We plan to create a teleconferencing network someday so we can reach all 600 Coors distributors at one time. We're also seriously considering videodiscs for sales, management, skills and safety training. It's what we'll all be doing soon."

BANK OF AMERICA

Applications

Bob Ripley has worked for the Bank of America in San Francisco for over ten years and notices a change in the way videotape training programs are used in his company:

> When you introduce a new banking service, it's actually a marketing program in many respects, because you're telling the employee what the new service is all about. But at the same time, you may be instructing employees on how to deal with customers' questions. And sometimes you're communicating with employees in the same production. It's a gray area because the finished program is not just a marketing presentation or a training videotape or a corporate communications tape. Some productions are straight marketing, some are straight training and some are straight corporate communi-

cation, but most we do are a combination of all of those
things.

The banking business has seen some radical shifts in the mid '80s.
Ripley has to keep up with those changes.

> We're staying busier and busier, and that is because of the
> banking environment right now. There's a lot of competition,
> a lot of reorganization. The bank is becoming more
> entrepreneurial. They're marketing more and more.

Thus, Ripley says, his department has to be more open to a wider
range of program needs. "We don't have one division in the bank that
dominates our time," he points out. "We're getting similar amounts of
work from all divisions."

Training is still a very important part of Bank of America's use of
video. But Ripley sees change coming.

> I don't see training as dominant anymore. It's still very
> important, and it's still going to be important. But one of
> the problems with training is that it's so volatile, depending
> on the corporation's point of history. If budgets get cut,
> training suffers.

So it's back to those gray area productions where more than one
goal is involved.

Ripley has done his part to save the bank money. The use of a training
videotape has cut back on travel expenses. "We've got 2000 offices spread
out across the country," Ripley says. "If people need to be updated
or trained on a particular subject, we use video training tapes. We do
training with video self-instruction on the use of the bank's computer
system."

Bank of America can reach up to 80,000 employees worldwide with
videotape. Ripley says the average audience is around 50,000. That is
a great deal of travel time saved through the use of video.

Another area of great importance is answering charges against the
banking industry in general and Bank of America in particular:

The bank is under attack from the press and from a lot of different areas. We're able to turn around a program, from a communications standpoint, answering these things in a very short period of time. From the time of inception to the time one of these programs is in every office can be four days or less.

Speed is important in order to reach all employees with the same consistent message.

Equipment and Staff

Bank of America's Media Services Department is located on two floors of an office building in San Francisco. The facility comprises two studios, two separate control rooms and full on- and off-line editing capability. All editing is done on five 1-inch machines. The department also has two 1-inch remote production packages. Distribution is done on 3/4-inch tapes.

The facility employs 25 full-time people. These regulars are supplemented by outside free-lancers, which can mean as many as 40 additional people.

Ripley believes that interactive video will soon have even greater impact.

We started it five years ago, but it is not big in the corporation. We did pilot programs, but we're not using it now. We try to push it as hard as we can, but we just don't have the say-so. We're a service and don't initiate productions. But if a department has the budget and wants an interactive program, we'll do it. It's still an expensive way to go with a lot of hardware to invest in, and with our 2000 offices you're talking about a huge expense.

Interactive will probably grow slowly. In other organizations it has a higher priority, but those companies don't have our distribution problem. We are also a service-oriented company without the manufacturing bottom line to work with. We don't have a widget to sell, just a service, and our people are the important links in making it work.

We have to train them and communicate with them and make
sure they understand what we're marketing.

BELK STORES

Jim Barefoot, Belk's manager of A/V Media Resources, remembers
that the Media Services Department came about in 1980 after individual
stores showed interest in video training:

> Stores started calling us. Our senior vice president challenged
> the training department and A/V staff; if we could supply
> the stores with video training programs that would better
> communicate with sales associates, increase their sales
> productivity and expand their fashion awareness, Belk might
> be persuaded to build a studio.

That challenge was met and exceeded. In the past the training and
A/V departments had produced a package of six to eight slide-sound
programs a year. Now it was time to move into video training.

Applications

In the first year of production Barefoot's department turned out 23
training videotapes. That number quickly climbed to 75 a year. Also
on the rise was the number of stores using the video training. After
the first year 65 stores were involved. By 1985, 240 of the 350 Belk
stores used video training, each with its own 3/4-inch playback system.
The stores are not required to buy the tapes that Barefoot produces,
but they do. The purchase of the tapes by the stores reflects a sincere
commitment to video. The training tapes resulted in increased sales and
better communications for the company.

The company has extensive printed materials on what videotapes are
available, and a guide to the company's procedures on producing a
videotape. It includes items on what type of program really needs to
be produced, what is involved with production, what the cost is, what
plan to follow in putting it all together and what to wear the day of
production.

Tapes that are produced fall into one of five categories. The first one deals with merchandise training and information, such as "Selling Glassware" and "Selling Better Men's Sportswear."

The second category is "How to Sell." "Selling Cosmetics," for instance, illustrates how to become a professional beauty advisor and work with the customer.

Category 3, "Skills-Training," covers items like the operation of store cash registers, "Effective Use of Stock Adjustment Sheets" and a four-program series on "Optical Character Recognition."

The fourth category includes meetings, motivational speakers, fashion shows or special guests.

Point of sale videotapes make up the last category. Barefoot even produced a companion tape on how to use "P-O-S" tapes effectively.

Jim Barefoot has found a new demand for video production outside of training itself. Stores have stepped up requests for the "P-O-S" tapes like "Fall '85 Horizon Collection of Creative Sewing" or "Chaus Fashion Show Fall/Winter '85." Both tapes run ten minutes and cost each store $20 and $25 respectively. To keep costs down on the "P-O-S" tapes, Belk Stores co-produces some of the programs with vendors. Barefoot is still making tapes in the other four categories but not in the numbers he once did. (Fig. 10.2)

Equipment and Staff

An initial investment of $120,000 was needed to give Belk Stores, the quiet giant of southern retailing, the studio and facilities it required. With more than 350 stores in 16 states, Belk employs close to 35,000 people. To keep them trained and informed the company created the Media Services Department mentioned above.

Barefoot sees a consistent growth in video for training and merchandising at Belk. "We're going to have video permanently installed in three different locations in every store. We already have stores that have bought a second and third machine. I think there will be one machine permanently installed in the store's conference room, where training classes are held and buyers and managers can watch corporate communication tapes. Another machine will be attached to a computer. Most stores already have a personal computer, and they will call up programmed instruction items in an interactive system. The third machine will be out in the store itself continually showing point of sale tapes."

Fig. 10.2. Belk Stores successfully uses video in all of its stores through point-of-purchase display consoles.

TUCSON POLICE DEPARTMENT

Applications

Jim O'Hara's responsibility as head of the Tucson Police Department's Video Production Unit is to use video to train over 650 police officers on everything from proper procedures, new laws and new departmental programs to in-service refresher training that could conceivably save a police officer's life. Details make the big difference. As O'Hara sees it, the overall philosophy of training video is, "Here's some information we want you to have for your own good. We care about you."

O'Hara says he considers his job no different from that of a video training producer working in any other corporate setting.

> When I meet with other people who do industrial video, for some reason, when they hear that I do law enforcement video, they say, "Well you're not doing what we're doing." But in fact I am doing the same thing. The expectation may be that any one member of my audience, if mistrained, could be the subject of a lawsuit against the city.

This adds to O'Hara's responsibilities and challenges.

O'Hara's involvement began eight years ago when, as an in-service training officer, he determined what it really cost the department to get needed training information to the line officers. "It was a two, three, four-week project to get out training materials and very man-hour expensive. I figured it costs us anywhere from $5000 to $7500 to put out 15 minutes of in-service training to the entire department." And because of staggered shifts, it took four weeks to reach everyone on staff. "We were real lucky to hit an 80% coverage because of days off and sick time," he says. "Now I put out a videotape and at the end of three days the entire department has seen it. Anyone who was on sick leave watches the video when he or she gets back. Now, if a new law goes into effect tomorrow, and I put out a training tape tonight, we can be assured that everybody who hits the street will be trained on that law."

As for saving costs, O'Hara says: "I can put out a videotape for anywhere from $500 to $2000. So I'm talking about a great deal less expenditure with more coverage and better retention. Instead of talking about how to stop a felony, I can show them."

Some important legal aspects of law enforcement training must be considered. O'Hara says video takes care of those problems:

> We have a permanent record of exactly what was taught. If I teach the same class 20 times, I can guarantee that I won't say the exact same words every time. All I've got to go on is a lesson plan, and that's usually in an outline form. Now we can go back and play the videotape and know precisely what an officer saw.

A new series of tapes just started by the department is called "Reminders." O'Hara explains:

> This is information that they already know, like "How to Approach a Doorway." Everybody knows how to approach a door properly, but after you've approached the doorway 3000 times, you can get lax. So we use the video training tape as a reminder. We say, "You already know this, but we're telling you again because it could save your life."

Another "Reminder" tape is called "Jamming Problems with a Smith & Wesson Revolver." The tape demonstrates, step by step, how to solve the problem quickly. Just after release it was put to the test.

"We had three situations in which officers were involved in shooting situations," says O'Hara. "The guns did jam and instead of panicking, the officers remembered the instructions on the tape." O'Hara produces at least one reminder tape a month.

Briefing tapes are kept to a maximum of 15 minutes. But O'Hara thinks 8 to 10 minutes is the best length because of attention span and retention. Officers see the tapes as part of their regular shift change briefings at roll call.

> We make videotapes to be used as training aids for what we call in-service training, which involves all police officers spending eight hours at the Training Academy three times a year. We cover items we couldn't cover in a 15-minute briefing tape. We use videotapes as training aids, but not

as the entire course. We also use video basic training to introduce a new skill to be learned.

The Tucson Police Department has put a high priority on video, making it second only to buying new police cars in the overall operating budget. Training is the primary use of video, but O'Hara says he also uses it as a communications tool and a morale booster. And occasionally video is used tactically—e.g., in prostitution decoy situations, for videotaping meetings with confidential informants and for taping major crime scenes.

Equipment and Staff

O'Hara is a one-man band. Other training officers are brought in to help from time to time but, for the most part, it's a single-person department. An average of three to four tapes a month are produced. Equipment is basically industrial standard. O'Hara uses a JVC camera and a Sony 5800 series editing system with a Convergence editor. Video training was started eight years ago with a Sony reel-to-reel tape deck and a black and white camera.

The Tucson Police Department turned to video training when it needed to solve a deadly and expensive problem—collisions between other traffic and police cars making Code 3 runs (at high speed, with lights flashing and sirens sounding). O'Hara went to work:

> I made a tape that outlined what the new Code 3 policy was.
> I included interviews with some of the officers who had been
> involved in fatal collisions. The reaction to the tape was almost
> 100% positive, and with a police audience, that's real tough.
> It's hard to get them to change their behavior.

In the four months after the tape was produced, Code 3 collisions were cut in half.

The push for more realistic training situations is constant. O'Hara says one standard shooting range procedure was changed after supervisors viewed a routine target practice taping. For range safety purposes an officer fired at a target then reholstered the empty gun and moved to the next target before reloading. The procedure was changed and a tape was made to emphasize the need to reload before putting a weapon back in a holster. It's a small change, but it creates a habit that carries over into real-life situations. Now before an officer walks out to the firing line, he or she sees a tape that explains all new procedures and the reasons for the changes. It saves time, and in the long run O'Hara believes it could save lives.

"There were no arguments about the changes," he says, "because everyone knew why they were made and even though we changed a procedure that had been standard for 10 or 15 years, everyone was able to adapt quickly and easily."

O'Hara hopes to work for an increase in realism and role-playing for the future, and he is pushing for interactive video. "The department is looking at the expense of hardware and the design of the programs. Interactive is not something I could whip out today and have in the field in a week. But we will be getting into it in the not-too-distant future."

TUCSON MEDICAL CENTER

Applications

The Tucson Medical Center (TMC) was founded in 1926 as a tuberculosis sanatorium. From the beginning the emphasis was on new techniques and the best in patient care. The hospital's Educational Services Department was formed in 1967. It grew over the years and now handles new employee orientation and an extensive in-service training program. Jerry Freund is Director of Education and has watched the department grow from doing simple classroom presentations to a heavy commitment to videotape training. "We did our groundwork and we have people around here more video-oriented than most organizations. Our people now think video," says Freund.

TMC spends from $60,000 to $80,000 a year on video production: e.g., "Demonstrating Surgical Techniques," "New Employee Orientation," "Management Development Skills" and "Improved Speaking Skills." The hospital also uses video for patient education via a "Master Antenna System" fed to every patient room.

Freund offers some specific guidelines on the use of video in training:

> Most video programs are designed to be utilized with an instructor. The typical sequence is to present four to five minutes of conceptual information on videotape where video can clarify the process being reviewed. The tape is then stopped

for group discussions. Examples include a behavior modeling sequence for teaching management development techniques. These are difficult to produce correctly. Another example is a tape on speaking skills, which is presented as part of a communication skills workshop.

Freund says other video programs are designed to stand alone. Examples include "Wellness and Patient Education" and the TMC-TV News program, which doubles as an employee training and communications program.

A less formal and more one-on-one video training system is located in the Learning Resource Center which is part of the hospital's medical library:

Copies of nonconfidential training tapes are filed in the medical library, where employees, students and physicians are encouraged to use the tapes in their free time. The programs have an average library shelf life of one year, but some, like tapes on surgical techniques, are timeless.

Also in the medical library is a system for physician continuing education. These programs are taped from the hospital's satellite downlink and played back at prescheduled times. Physicians can sign the tapes out for viewing at home.

Freund says the flexibility of TMC's distribution system makes a big difference in getting the message across. Besides the "Master Antenna System" and the satellite time shift recording, the hospital uses a number of 3/4-inch videotape machines and monitors on carts that are assigned to various locations throughout the facility.

This flexibility is also evident in the content of the tapes. In many regular TMC training tapes humor plays a key role. Employees might see a take-off on the old *Sea Hunt* series that stresses on-the-job safety, or an impersonation of Humphrey Bogart in *Casablanca* describing proper fire evacuation procedures. One *Magnum P.I.* take-off designed to get employees involved in community meals-on-wheels programs was re-edited and is now shown as a public service announcement on commercial television in Tucson.

Another series of tapes produced regularly is called "Trigger Tapes."

They're used to stimulate discussion around an idea or procedure. This is a "How would you handle this?" training situation on tape. It's used to reach certain hospital procedures and is used in a classroom with an instructor.

Equipment and Staff

TMC's video facility consists of an office/editing area with several classrooms adapted for live or taped productions. Equipment includes a Sony 5800 Series editing system and two Sony 1800 cameras with portable recorders. The hospital also has a direct tie into Tucson's community access cable television system. Health and preventive medicine programs are produced and aired on cable. The cable tie-in is part of TMC's Community Health and Wellness Program. The program also allows patients to sign out 1/2-inch videotapes for home viewing.

Another key to success, according to Freund, is awareness.

"The training staff has been sensitized to creative applications and appropriate uses of video," he says. "Not every idea suggested is produced." Freund says once the decision to produce a tape is made, there is still a great deal of time spent on "front-end analysis" of how it will be used:

Sometimes there exists a need for the "talking head" approach. Not every training video production needs to be an award-winning program. The bottom line is "is it appropriate and do you get the message across?" Production values can be compromised at times.

Freund stresses the fact that there is a limit to those compromises, and that the audience must be taken into consideration when making production decisions.

In the future Freund sees more growth in programs received by the hospital satellite dish: "The greatest growth has been in the application of satellite video conferencing services. The cost of satellite delivered training programs can become as low as $1 per viewer."

He also sees potential for increased marketing opportunities for hospital services and increased participation in live hospital video networks. "The hospital's commitment to satellite technology is already paying off and will continue to do so at an accelerated rate," Freund concludes.

Freund also sees a day when learning and the hardware used in teaching get closer together.

> I really see a lot of exciting developments where the student can be very flexible with his or her style or way of learning. We're already seeing major changes in educational technology, especially in computer-assisted instruction, where there's a great deal of flexibility.

Freund foresees a new technology that could simplify further the handling of video training:

> As an educator, I'm looking at what is effective for distribution, and what I think will take the world by storm is 8mm video. I've been debating just how far to go with 1/2-inch tape because I'm still fighting the format battle (VHS and Beta). I think that we will see 8mm as a major in-house distribution format for video training.

Freund cites the small size, the cost of the cassette, flexibility and the lower cost of playback equipment as primary reasons for his consideration of changing all of TMC's training to 8mm.

> The cost will allow us to do away with the roaming AV carts (currently 3/4-inch machines and monitors), and replace them with permanent 8mm playback installations. It just makes good business sense and is much more cost-effective. I can foresee doing employee orientation on 8mm as part of the package the new person takes home. I can see more and more continuing education being done in the home. I also see 8mm used in videotaping patients, particularly in the rehab area, creating tapes that show the patients what they have to do when they go home.

Farther in the future Freund sees a special application for high definition television (HDTV) at TMC. "One of the limitations with the current NTSC standard is the loss of details that can be critical in a surgical technique," he says. "Right now the cost is prohibitive, but I think that that will change soon."

Freund also says the higher definition is needed for fully using a new diagnostic technique called Magnetic Resonance Imaging (MRI), the next step in CAT scans.

> A patient may come here to Tucson Medical Center to have an MRI done. But the doctor could use a closed circuit to call up the scan on a HDTV screen at any of the other hospitals in Tucson. It's a project I'm working on right now. It obviously has great educational potential. The detail is so important and the scans could be used for teaching.

Afterword

I hope this book has provided a useful guide to the design and production of video training programs. It should become clear in going through the phases of program development how important it is to have clear objectives for the training event and how good planning will affect the end result.

Perhaps the most significant phase in the development of video training programs is the design phase. I hope this book emphasizes this phase. It is through careful objective setting and thorough audience analysis that the video training program becomes successful. Whether we are developing linear programs or interactive programs, the design phase is critical. The choices we make and the objectives we select will influence the rest of the production process.

The production of video training programs as outlined in this book is a complex series of events, all intertwined to create an end result that causes a change. Each step along the way is important, and each step will be affected by the training objectives *and* the amount of effort placed on the planning stages. The planning of the program (Chapters 6, 7 and 8) will definitely influence the outcome. I believe that the success of the video training program will be proportionate to the amount of planning that takes place.

Training will take us into the future. As organizations move into the 21st Century, and as the technology required to move those organizations advances, more emphasis will be placed on training to prepare workers for these changes.

Video will continue to grow along with training and provide an effective delivery tool. Computer-enhanced video will become the dominant training method. As costs decrease and the demand for individualized training increases, the computer linked with videodisc will appear in more and more organizations. The importance of training workers to change and grow with technology will create the budgets necessary to support computer-enhanced video.

This demand for advanced, individualized training will create a need for training teams that are specialized in instructional design principles, computer programming and video production. The blend of computers, video and education will produce training results that will prepare people for the future.

Emphasis will be placed on training results and production values that reflect the resources committed to create change in the work place. It is our responsibility as video producers and trainers to stay abreast of changing video technology and training techniques so that we can increase values and provide cost-effective programs that create the desired change.

Appendix

The following chart, which roughly parallels the discussion in chapters 4-8, is used by the author in his seminars.

VIDEO TRAINING PROGRAM DEVELOPMENT

PHASE I — DESIGN

STEP	TASK	SUB TASKS	AGENT
1.	Problem Analysis	Determine the problem	Content Expert
		Determine cost to organization	Instructional Designer
		Determine evaluation system	Video Producer
2.	Audience Analysis	Determine audience profile	Content Expert
		Assess audience needs	Instructional Designer
		Assess audience delivery system	Video Producer
3.	Objective Setting	Write program objectives	Instructional Designer
		Determine content sequence	
		Determine total number of instructional modules and segments	
4.	Learning Strategy	Develop learning strategy	Instructional Designer
		Determine training delivery method	
5.	Media Selection	Determine resources, budget and time frame	Instructional Designer Video Producer
		Select media	
6.	Program Production Schedule Development	Estimated time required to complete program	Instructional Designer Video Producer

PHASE 2
SCRIPTING AND STORYBOARDING

7.	Research	Gather current content available	Content Expert
		Determine media available	Instructional Designer
8.	Treatment Development	Determine approach	Video Producer
9.	Program Outline Development	Start developing and sequencing content Check outline against objectives	Instructional Designer

SCRIPT DEVELOPMENT

10.	Rough Script Development	Write rough script, expanding approach	Content Expert Video Producer
11.	Printed Materials Development	Start to prepare workbooks, tests and all printed support materials	Instructional Designer
12.	Review of Rough Script	Make comments and corrections	Content Expert Instructional Designer Video Producer
13.	Writing Second Script Draft	Incorporate recommended changes	Video Producer
14.	Storyboard Development	Visualize rough script	Video Producer
15.	Review of Second Script, Storyboard and Printed Materials	Make comments and corrections	Instructional Designer Content Expert Video Producer
16.	Writing Final Script, Storyboard and Printed Materials	Incorporate recommended changes	Instructional Designer Video Producer
17.	Review of Final Script Storyboard	Sign-off approvals, deliver to production staff	Content Expert Instructional Designer Video Producer

PHASE 3
PRE-PRODUCTION

18.	Develop Shooting Script	Divide script into scenes Group scenes by location Group shots within scenes Determine camera angles Determine camera movement Create audio cues Time scenes and shots Create transitions where needed	Video Producer Production Assistant
19.	Determine Production Requirement	Identify graphic requirements, props, sets, special logistic considerations Determine locations Determine personnel needed Determine talent requirements Determine technical requirements Determine audio and lighting requirements	Video Producer
20.	Develop Production and Post-Production Schedule	Schedule production time Schedule studio and location time Schedule post-production time	Video Producer
21.	Graphic Support Meeting	Discuss preliminary graphic support, design special effects and start design process	Video Producer Graphic Artist
22.	Begin Production of Graphics		Graphic Artist
23.	Scout Locations	Evaluate locations for special requirements, angle and placement	Video Producer Production Assistant

24.	If Sets Are Required:	Design and construct sets	Production Assistant
25.	If Talent Is Required:	Audition and hire talent, write contracts, send script, schedule	Production Assistant Video Producer
26.	Order Required Materials	Order videotape, audiotape, special equipment	Production Assistant
27.	Create Shot Sheet	Based on location and script, create shot sheet and distribute	Video Producer
28.	Equipment Arrangements	Rent equipment, hire crew	Video Producer

PHASE 4
VIDEO PRODUCTION

	Record all Location Shots	Direct transportation, staging, personnel, facilities, equipment, rentals, meals	Video Producer Production Assistant Content Expert
30.	Record all Studio and Insert Shots	Direct talent, graphics special effects	Video Producer Production Assistant (Content Expert)
31.	Revise Script According to Location Shoot	Prepare for audio narration recording	Video Producer
32.	Record Audio Voice over Narration	Direct all audio recording	Video Producer

PHASE 5
POST-PRODUCTION

33.	Log and Review Tapes	Make window dubs if time code necessary. Log and review tapes for notes and accuracy, time code log	Production Assistant
34.	Create Edit Script	Revise script for editing	Video Producer
35.	Music and Sound Effects Selection	Select all music and stock sound effects	Production Assistant
36.	Record Music and Sound Effects	Prepare tracks, mixes, recording of audio, including narration mix	Production Assistant (Video Producer)
37.	Rough Edit	Do paper edit (offline) of the coded tapes, perform rough edit	Video Producer
38.	Review of Rough Edit	Note revisions, check for accuracy	Video Producer Content Expert
39.	Re-Edit	Incorporate revisions	Video Producer
40.	Final Edit	Add inserts, graphics, audio mix	Video Producer
41.	Review of Final Complete Programs	Review videotape, printed materials	Educational Designer Content Expert Video Producer
42.	Distribution	Prepare printed materials and videotapes for distribution	Video Producer Production Assistant

Bibliography

TRAINING

Anderson, Ronald H. *Selecting and Developing Media for Instruction.* New York: Van Nostrand Rheinhold, 1976.

Craig, Robert L. *Training and Development Handbook.* 2nd ed. New York: McGraw-Hill, 1976.

Gagne, Robert M. and Leslie J. Briggs. *Principles of Instructional Design.* New York: Holt, Rinehart & Winston, 1979.

Instructional Design: A Plan for Unit and Course Development. 2nd ed. Belmont, CA: Fearon Publishers, 1977.

Kemp, Jerrold E. *Planning and Producing Audiovisual Materials.* New York: Harper & Row, 1980.

Laird, Dugan. *Approaches to Training and Development.* Reading, MA: Addison-Wesley, 1978.

Nadler, Leonard. *Designing Training Programs: The Critical Events Model.* Reading, MA: Addison-Wesley, 1982.

Preparing Instuctional Objectives. Revised 2nd ed. Belmont, CA: Fearon-Pitman, 1984.

Training and Technology: A Handbook for HRD Professionals. Reading, MA: Addison-Wesley, 1984.

Warren, Malcolm W. *Training for Results.* Reading, MA: Addison-Wesley, 1979.

VIDEO PRODUCTION

Anderson, Gary. *Video Editing and Post-Production: A Professional Guide.* White Plains, NY: Knowledge Industry Publications, Inc., 1984.

Directing, Getting into Video, Introduction to Electronic Editing, Scriptwriting, and similar titles are available from Mattingly Publications, Fairfax, VA.

Hickman, Walter A. and Milan Merhar. *Time Code Handbook.* Boston, MA: Cipher Digital, 1982.

Marlow, Eugene. *Managing the Corporate Media Center.* White Plains, NY: Knowledge Industry Publications, Inc., 1981.

McQuillin, Lon. *The Video Production Guide.* Indianapolis, IN: Howard W. Sams & Co., 1983.

Millerson, Gerald. *The Technique of Television Production.* London and Boston: Focal Press, 1979.

Van Deusen, Richard E. *Practical AV/Video Budgeting.* White Plains, NY: Knowledge Industry Publications, Inc., 1984.

Van Nostran, William. *The Nonbroadcast Television Writer's Handbook.* White Plains, NY: Knowledge Industry Publications, Inc., 1983.

Wiegand, Ingrid. *Professional Video Production.* White Plains, NY: Knowledge Industry Publications, Inc., 1985.

Zettl, Herbert. *Sight, Sound, Motion: Applied Media Aesthetics.* Belmont, CA: Wadsworth Publishing Co., 1973.

— — —. *Television Production Handbook.* 4th ed. Belmont, CA: Wadsworth Publishing Co., 1984.

INTERACTIVE VIDEO AND COMPUTER-BASED TRAINING

Decker, Walker F. and Robert D. Hess. *Instructional Software/Principles and Prospectives for Design and Use.* Belmont, CA: Wadsworth Publishing Co., 1984.

Flowcharting Guidelines. Montvale, NJ: Pioneer Video, Inc., 1984.

Floyd, Steve and Beth Floyd. *Handbook of Interactive Video.* White Plains, NY: Knowledge Industry Publications, Inc., 1982.

Heines, Jesse M. *Screen Design Strategies for Computer-Assisted Instruction.* Bedford, MA: Digital Press, 1984.

Iuppa, Nicholas V. *A Practical Guide to Interactive Video Design.* White Plains, NY: Knowledge Industry Publications, Inc., 1984.

Post-Production and Formatting Information. Montvale, NJ: Pioneer Video, Inc., 1984.

Scheduling, Authoring, and Project Management Guide. Montvale, NJ: Pioneer Video, Inc., 1984.

Steinberg, Esther R. *Teaching Computers to Teach.* Hillsdale, NJ: Lawrence Erlbaum Associates, 1984.

Index

About the Author

Steve R. Cartwright holds a degree in communications and for over fifteen years has applied a unique background that combines skills in both training program design and video production to creating effective video training programs. As a trainer and video producer, Mr. Cartwright has designed, written and produced training programs for health care, government, law enforcement and industry.

As Manager of Media Services for Burr-Brown Corporation in Tucson, AZ, Mr. Cartwright produces technical training programs for the microelectronics industry and is responsible for producing corporate communications, management training and sales training.

Mr. Cartwright is a recognized expert in the video training field, and through his numerous workshops and seminars he teaches others how to design, write and produce effective video training programs. As Program Director for Video Expo and North American Television Institute, Mr. Cartwright designs and organizes seminars in video design, production and management for Knowledge Industry Publications, Inc. He is a member of the Association of Educational Communications and Technology and the International Television Association.